Neuroethics:

Mapping The Field

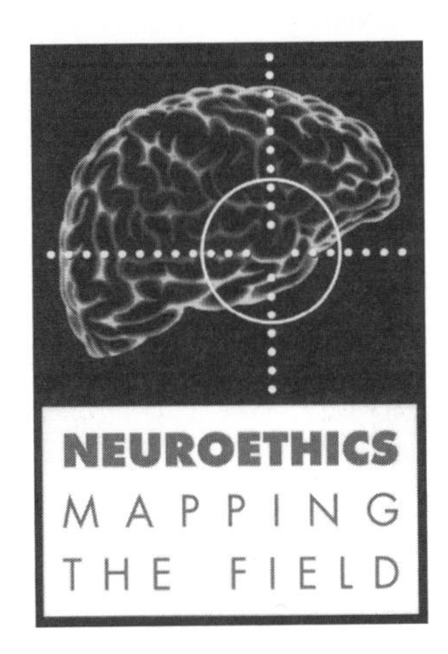

Conference Proceedings

May 13-14, 2002
San Francisco, California

Sponsored by:
The Dana Foundation

Hosted by:
Stanford University
and
University of California, San Francisco

Steven J. Marcus, Editor

Published by The Dana Press
J. Andrew Cocke, Project Editor
Randy Talley, Production Manager
Designed by Renée Roberts
ISBN 0-9723830-0-X

Neuroethics : mapping the field : conference proceedings, May 13-14, 2002, San Francisco, California/sponsored by the Dana Foundation; hosted by Stanford University and University of California, San Francisco ; Steven Marcus, editor.
p. ; cm.
Includes bibliographical references and index.
ISBN 0-9723830-0-X (pbk. : alk. paper)
1. Neurosciences--Moral and ethical aspects--Congresses. 2. Neurology--Moral and ethical aspects--Congresses.
[DNLM: 1. Bioethical Issues--Congresses. 2. Neurology--standards--Congresses. 3. Neurosciences--standards--Congresses. 4. Individuality--Congresses. 5. Personhood--Congresses. 6. Social Responsibility--Congresses. WL 21 N494 2002] I. Marcus, Steven, 1942- II. Charles A. Dana Foundation.
RC327 .N36 2002
174'.957--dc21

2002151389

The Dana Foundation
745 Fifth Avenue Suite 900
New York, NY 10151

Also available in PDF at www.dana.org

Editor's note

On May 13-14, 2002 more than 150 neuroscientists, bioethicists, doctors of psychiatry and psychology, philosophers, and professors of law and public policy came together in San Francisco, California to participate in a landmark conference *Neuroethics: Mapping the Field.* The conference organizers define "neuroethics" as the study of the ethical, legal amd social questions that arise when scientific findings about the brain are carried into medical practice, legal interpretations and health and social policy. These findings are occurring in fields such as genetics, brain imaging, disease diagnosis and prediction. Neuroethics should examine how doctors, judges and lawyers, insurance executives and policy makers as well as the public will deal with them.

The seven formal sessions and the floor discussions following each session have been edited for readability. Biographical notes on the speakers are included in an appendix of this volume, as well as the members of the conference planning committee. This book is also available in PDF format at www.dana.org

Neuroethics: Mapping The Field

Contents

Luncheon Speech

Session III

Ethics and the Practice of Brain Science

Dinner Speech

Session IV

Brain Science and Public Discourse

Session V

Mapping the Future of Neuroethics

Welcome

ZACH W. HALL: It's my very great pleasure to welcome you this morning on behalf of the Dana Foundation, UCSF, and Stanford, to this conference. "Neuroethics: Mapping the Field" is an attempt to bring together scientists, ethicists, humanists, and those concerned with social policy to reflect on the broad implications of current and ongoing research on the human brain.

This meeting had its genesis in a visit to San Francisco by Bill Safire about a year and a half ago. I took Bill down to the new Mission Bay campus at UCSF, and we were talking about all the brain research that would be going on there. I said that we also hoped to have a

Dr. Zach Hall, University of California, San Francisco.

bioethics center. As we were talking about the need for discussion of these issues with respect to the brain, Bill suddenly turned to me and said, *neuroethics*. It was like that magic moment—"plastics," in the movie *The Graduate*. Bill said, "neuroethics," and I thought, "that's it."

It was a recognition that these problems are so serious and have such broad implications that they deserve a special designation. And, indeed, out of that came a special meeting to consider them. Let me make a couple of comments about the sort of spirit that has guided the organization of this meeting. First of all, we have tried to bring together people from different disciplines, and we have tried to maintain a balance among the different fields in each panel.

Secondly, this is a meeting for discussion. It's not a meeting in which the audience is going to sit and be talked at, and we have invited a number of distinguished people to the meeting who are not speakers, but whose commentary, and whose presence, and whose ideas will enrich the meeting. And, finally, this is not a meeting to give a final answer to any question. It is really a meeting to raise questions, and to sort of lay out what we see as a terrain whose exploration will surely take many years to pass.

Introduction

Visions for a New Field of "Neuroethics"

William Safire
Chairman,
The Dana Foundation,
and columnist,
The New York Times

The first conference or meeting on this general subject was held back in the summer of 1816 in a cottage on Lake Geneva. Present were a couple of world-class poets, their mistresses, and their doctor. They'd been reading and discussing the disturbing works of Erasmus Darwin—who later had a grandson named Charles Darwin—about the creation of artificial life.

It was near the end of the Enlightenment, the era when, with the world in an intellectual revolt against the despotism of kings and the power of the clergy, a philosophy of rationalism and tolerance had burst upon the scene and, with it, political revolution in America and in France. Thinkers wrote about the perfectibility of man. Minds had been opened, morality reexamined, even as some of the reformers brought

William Safire, Chairman, The Dana Foundation.

on their own reign of terror, and a conservative reaction was setting in.

One of the poets at that lakeside gathering, Lord Byron, had a bright idea to enliven the discussion. "Let's each of us write a ghost story," he suggested. He tried and couldn't get started. His friend, Percy Bysshe Shelley, also had a go, but quickly set it aside. Their doctor came up with a sorry tale about a vampire. And Byron's mistress only wanted Byron.

The young woman with Shelley, however, was caught up in the terror of the manipulation of life by the new science. She was the strong-minded daughter of Mary Wollstonecraft, the pioneer feminist and moral rebel. And her father was William Godwin, the social philosopher and anarchist. She wrote her ghost story and married her poet. Two years later, Mary Shelley's *Frankenstein: The Modern Prometheus* was published. Prometheus, you remember, was the god who was tortured for all eternity for bringing to man godlike powers.

In our time, two centuries later, man's Promethean presumption to create life, to interfere with what had been the exclusive domain of God or nature, is being fiercely debated all around the world. Europe is consumed with controversy about the genetic modification of foods, and, with a nod to Mary's monstrous creation, the improved—or at least manipulated—products are derided as "Frankenfoods." The fear of playing God, as well as the countervailing hope of creating lifesaving life in the laboratory, roils the public reaction to science's breakthroughs in our own new enlightenment.

Welcome to the first symposium on one specific portion of that two-century-long growing concern: neuroethics—the examination of what is right and wrong, good and bad about the treatment of, perfection of, or unwelcome invasion of and worrisome manipulation of the human brain.

The fear of playing God, as well as the countervailing hope of creating lifesaving life in the laboratory, roils the public reaction to science's breakthroughs in our own new enlightenment.

It's fitting that The Dana Foundation be the conference's sponsor. For the past decade we've been focused on the brain, not only by directly funding researchers in many fields of neuroscience—from brain imaging to neuroimmunology—but also by marshaling other support, both private and public, for brain research.

The Dana Alliance for Brain Initiatives in the United States and the European Dana Alliance for the Brain are a network of more than 300 leading neuroscientists, including more than a dozen Nobel laureates, actively reaching out to explain their work and offer help to the general public. It has been successful, and this fall in Washington, D.C., and next year in London we'll be opening centers to let more scientists, philosophers, critics, and even newspaper pundits engage in informed discussions.

One of our founding alliance members and driving forces, Zach Hall, suggested that the time was right for this conference in this place. And Dana Foundation president Ed Rover, executive vice president Francis Harper, and I were pleased to work with

Barbara Koenig and Judy Illes at Stanford to help make it happen.

The only field in which I can claim some expertise in this crowd is the English language. I'm a pop grammarian and etymologist, and I regularly get asked questions like "Where does 'the whole nine yards' come from?" It comes from the capacity (measured in cubic yards) of a cement truck. I once wrote that, and I got a bunch of letters back from people saying, "It's not a cement truck, it's a *concrete* truck."

Another question I regularly get asked is, "What's the difference between ethics and morals?" The Latin *moralis* was formed by Cicero as a rendering of the Greek *ethikos*, and the words have been used interchangeably ever since. But in their usage a distinction can be drawn. To me, *moral* has to do with right and wrong, and *ethics* with good and bad.

Now, what's right is good and what's wrong is bad, so there's a lot of overlap. But I think of the difference this way: *moral* implies conformity to long-established codes of conduct set primarily by religious authorities, while *ethical* involves more subtle questions of equity. The moralist asks, "Is it right by intrinsic standards?" The ethicist asks, "Is it fair in the light of this society's customs and in these times?" Moral connotes standing firm; ethical, while still pretty stiff, can be said to swing a little.

Neuroethics, in my lexicon, is a distinct portion of bioethics, which is the consideration of good and bad consequences in medical practice and biological research. But the specific ethics of brain science hits home as research on no other organ does. It deals with our consciousness—our sense of self—and as such is central to our being. What distinguishes us from each other beyond our looks? The answer: our personalities and behavior. And these are the characteristics that brain science will soon be able to change in significant ways.

Let's face it: one person's liver is pretty much like another's. Our brains, by contrast, give us our intelligence, integrity, curiosity, compassion, and—here's the most mysterious one—conscience. The brain is the organ of individuality.

Zach Hall has made the point that when we examine and manipulate the brain—unlike the liver or, as Art Caplan would have it, the pancreas—whether for research, treatment of disease, or perhaps-sinister political ends, we change people's lives in the most personal and powerful way. The misuse or abuse of this power, or the failure to make the most of it, raises ethical challenges unique to neuroscience. What's more, neuroscientists have a built-in conflict of interest that sets them apart from all other ethicists.

What distinguishes us from each other beyond our looks? The answer: our personalities and behavior.

Everybody's brain has a personal, selfish interest in the study of the brain. It is the ultimate in self-dealing. Won't a human brain tend to do what's best for itself and take charge and take chances, plunging ahead to treat or improve the brain, as the brain might not do for the same body's liver? In possession of this power of self-improvement, of "perfectibility," how will we define and protect the integrity of our ability to judge morally and conduct ourselves ethically?

I hope the proceedings today and tomorrow will

concentrate on the special challenges of neuroethics and not keep punching away at the bioethical hot buttons of embryonic stem cells and cloning so heavily debated elsewhere. Here are a few examples of the questions I hope we cover:

Remember the psychosurgeries for aggression some forty years ago? What ethical rules or legal regulations should there be for treatment to change criminal behavior?

Suppose we could develop a drug to make someone less shy, or more honest, or more intellectually attractive, with a nice sense of humor. What is there to stop us from using such a "Botox for the brain"? More seriously, if a person's brain is impaired by disease, injury, or mental illness, and he or she cannot give informed consent, who is to decide when participation in a clinical trial is humane and proper? Doctor, relative, researcher, insurer, or court?

Should we develop a drug to improve memory or to repress painful remembrances? Or to help a prosecutor elicit a professedly forgotten detail? Is it fair to implant a chip in the brain to enhance memory before an academic examination? Or is that like giving a steroid to an Olympic athlete? And here's one for the defenders of privacy: Is the imaging of suspected terrorists' brains to detect lying a form of torture, or at least a way of forcing people to incriminate themselves?

As we learn that memory is not fixed, but is constantly being reshaped as reminiscences are recalled and stored again, how do we even *define* truthful testimony and judge its reliability?

In discussions of ethics in every field there's a

"but what if?" factor that fuzzes clear lines. A doctor considers it an ethical responsibility to inform a patient of the seriousness of his or her illness, but what if the patient is depressive and a suicide risk? A geneticist may consider it ethical to warn a person of the likelihood of some great vulnerability, but what if that means the patient won't be able to get insurance?

A journalist considers it unethical to reveal a source who was promised confidentiality. But what if the source turns out to be lying, or the source has evidence to save an accused from jail?

I'd like to hear about some of these "but what if's" and other questions in neuroethics. The people in this room are better equipped than most to take them on, and they may proceed today and tomorrow to carve out new territory for an old philosophical discipline. This could well be a historic meeting that participants will look back on with great pride and that others will talk about as a seminal moment in the development of this new field.

I expect that a book of your conference papers will be published along with some of the lively and profound give-and-take. It won't have the sales of *Frankenstein*, and Boris Karloff won't star in the movie, but it might help, as you put it, map the field. Thank you.

Session

I

Brain Science and the Self

Albert R. Jonsen, Session Chair

Emeritus Professor of Ethics and Medicine, University of Washington

Antonio R. Damasio

Van Allan Professor and Head of Neurology, University of Iowa

Patricia Smith Churchland

President's Professor of Philosophy and Chair of the Philosophy Department, University of California, San Diego

Kenneth F. Schaffner

University Professor of Medical Humanities, George Washington University

Jonathan D. Moreno

Kornfeld Professor of Bioethics and Director of the Center for Biomedical Ethics, University of Virginia

ALBERT R. JONSEN: The aim of this first session is to explore some of the broad questions that underlie the relationship between ethics and the neurosciences.

It so happens, fortuitously and maybe a little eerily, that our conference takes place 142 years from the death, almost to the day, of Phineas Gage right here in San Francisco. Gage is a figure of some interest in the neurosciences. Thirteen years before his death, he was the victim of a freak accident. An iron bar was blasted into his left cheek, through the frontal lobe of his brain, and out the top of his skull. Gage lived with his physical capacities intact and his cognitive faculties unimpaired—though with one significant exception. He became incapable of making moral choices.

Gage's remarkable history has frequently been told, but never so appropri-

Dr. Albert Jonsen, University of Washington.

ately for the purposes of this conference than by our first speaker, Dr. Antonio Damasio, in his fine book *Descartes' Error: Emotion, Reason, and the Human Brain.* His wife, Dr. Hanna Damasio, who is here today, was a major figure in reconstructing the neuroanatomy of Gage's brain for purposes of modern study.

Dr. Antonio Damasio opens his book with Gage's story. He writes: "Gage had once known all he needed to know about making choices conducive to his betterment. He had a sense of personal and social responsibility. He was well adapted in terms of social convention and appears to have been ethical in his dealings. After the accident, he no longer showed respect for social convention, ethics were violated, the decisions he made did not take into account his best interests. There was no evidence of concern about his future, no sign of forethought."

Dr. Damasio then turns to a patient of his own, named Elliot, whom he calls a modern Phineas Gage. He describes Elliot and similar patients as powerless to proceed from understanding a moral situation to making a moral choice, and he explores their neuropsychology, neuroanatomy, and neuropathology, which seem to occasion this radical destruction of personality.

So it is fitting that this session begins with Dr. Damasio, who is Van Allen Professor and Head of

Neurology at the University of Iowa.

Our second speaker is Dr. Patricia Churchland, who is the University of California President's Professor of Philosophy and Chair of the Philosophy Department at the University of California, San Diego. Her most recent book is *Brain-Wise: Studies in Neurophilosophy.*

Our third speaker is Dr. Kenneth Schaffner, University Professor of Medical Humanities at George Washington University. His most recent book is *Discovery and Explanation in Biology and Medicine.*

And our final speaker is Professor Jonathan Moreno, Kornfeld Professor of Bioethics and Director of the Center for Biomedical Ethics at the University of Virginia, whose most recent book is *Undue Risk: Secret State Experiments on Humans.*

The Neural Basis of Social Behavior: Ethical Implications

SUMMARY: Dr. Damasio discussed the social and emotional foundations of ethics and pointed out that ethical behaviors are present not only in humans but also in other species. This indicates that ethics results in good measure from evolution, that it is another aspect of bioregulation. But he warned that there are no moral "centers" of the brain—though extensive neural systems are indeed involved—and that although genes impel our ethical behavior, they do not compel. Such behavior varies with our culture, our living situation, and the health of our brains.

ANTONIO R. DAMASIO: Ethical behaviors are a subset of *social* behaviors; it's not possible to conceive of ethics outside the concept of society. And because there are nonhuman societies, the essence of ethical behavior does not begin with humans. There is evidence from primates and other species—from vampire bats to wolves—of conduct that appears, to our cultivated eyes, as moral conduct. Altruism, censure, recompense for certain actions, and compassion are evident examples in nonhuman and even nonprimate species.

Moreover, because the expression of ethical

behavior is associated to a great extent with "social" emotions—such as sympathy, shame, embarrassment, guilt, and the form of social anger we call moral indignation—there is also evidence for such emotions not only in humans but in other species.

Human ethics has a degree of complexity that makes it distinctly human.

This realization—it is a fact, not a hypothesis—may come as a shock for those who believe that ethical behavior is a distinctive human trait. As if it were not enough to have Copernicus tell us we are not at the center of the universe, Darwin tell us that we have humble origins, and Freud tell us that we are not masters of our own house, now we are being told that even in the realm of ethics there is forerunner behavior.

For those who feel humiliated by this revelation, however, please note that human ethics has a degree of complexity that makes it distinctly human. The refinement is human; the codification we have enacted on ethical behaviors is human; the narratives we have built around the situation are obviously human. It is only the basic behaviors behind the situations that are not uniquely human.

Dr. Antonio Damasio, University of Iowa.

This fact—the relation of seemingly moral conduct with the emotions—has inspired in my work the following hypothesis:

The construction we call ethics began with the edifice of bioregulation. By *bioregulation* I mean the set of automated mechanisms that allows us to balance our metabolism, maintain life, and achieve well-being, and which also produces drives and motivations, emotions of diverse kinds, and feelings.

Note that I am not reducing ethics to a simple matter of evolution, or of gene transmission or expression, or of brain structures alone. As conscious, intelligent, and creative creatures inhabiting a cultural environment, we humans have been able to shape the rules of ethics, shape their codification into law, and shape the application of the law into what we call justice. And we continue to do so. In fact, one purpose of conferences like this is to discuss ways in which we may shape the rules of ethics in keeping with the new problems posed by advances in science and technology.

So ethics is not just about evolution, even if I am suggesting that it starts with evolution. And it is not just about the brain. Culture does the rest, and the rest may be most of it.

Similarly, elucidating the biological mechanisms underlying ethics does not mean that those mechanisms, or their dysfunction, ensure certain behaviors. There certainly are determinants of behavior that come from our evolutionary biology—from the way our brains get set, and from the ways they get set both by genes and by the culture in which we develop—but there is still a degree of freedom that allows an individual to intervene. As far as I can see, there is free will—though not for all behaviors, and not for all conditions, and sometimes not to the full extent in any condition.

Unsurprisingly, I believe that what we call ethics

Session I (left to right): Drs. Moreno, Schaffner, Churchland, Jonsen and Damasio.

today depends on the workings of certain brain systems. But now come a few additional disclaimers.

First of all, I am talking about *systems*, not centers. On a number of occasions I've pleaded with science writers not to talk about a brain "center" of anything—not of language, or memory, or morality—but the plea is often disregarded by the headline editor.

Talking of "centers" gives the false impression that there's some kind of clearinghouse in the brain, in charge of a certain set of behaviors. Nothing could be further from the truth. We are in fact dealing with systems made up of several components that maintain complex interactions among themselves. It is only when those systems operate, in a given context, that certain kinds of behaviors emerge, along with certain kinds of cognition related to those behaviors.

The second disclaimer is just as important:

Although certain systems in the brain are clearly related to moral behavior, they are not set by genes to operate for the purposes of morality and ethics. These systems are indeed dedicated, for example, to memory of particular kinds, or to decision making, or to creativity. But they respond to certain needs of an individual living in a social collective—to help the person harmonize with the conditions in which he or she is living—and these needs and conditions arise independently from what evolution has equipped us with.

The upshot is that ethics is a wonderful by-product. We could not have it if we did not have a capacity to learn, if we did not have a capacity to recall, if we did not have a capacity to imagine, reason, and create. But I doubt that there is a dedicated moral system in the brain, and certainly no moral center.

This is not to say that damage to the brain will not result in moral impairment. We now have a large collection of data on patients who are Gage-like in some way. They learn, they recall, they preserve their language, they manage logic quite nicely. And yet, even if they recall social conventions and rules of ethical behavior, they are no longer able to apply them effectively. Though they know what is "right" and "wrong" and "good" and "bad," they are impaired in a whole class of social emotions such as embarrassment, shame, guilt, and sympathy. This, in turn, impairs the decision-making mechanism that is needed for appropriate social management, and subsequently impairs any new learning of this sort of social knowledge.

The cases I am referring to are of adults who have been upstanding members of society up to the point when their brains sustained damage. What happens if the brain is damaged much earlier? Recently we reported the cases of two patients who had sustained lesions of the same sector of the brain but at very young ages—one of them in the second year of life, the other even before the end of the first year.

We were able to study these patients in their twenties. In many respects they were entirely comparable in their behaviors to those who had sustained their lesions in adulthood, but there was an important distinction: they had not been able to learn the social conventions and ethical rules the adults had learned. It was not just a matter of not acting on the rules—they had not learned the rules to start with. And this, predictably, had led to a much worse quality of behavior.

So I would like to close by posing a question. Just imagine that by a quirk of fate, evolution had gone in a different way and humanity had come into being with the kind of losses that the Gage-type patients have sustained, and that we would not have the possibility of expressing the social emotions. What would the world have been like? Would ethics have ever developed? Would we be here to tell?

Neuroconscience: Reflections on the Neural Basis of Morality

SUMMARY: Dr. Churchland discussed the proposition that the brain relies on "inner models" and "emulators" for promoting one's survival and well-being. In any given situation, these mechanisms simulate alternative options and predict their expected outcomes to enable an individual's decision making—that is, to take particular actions, or not to. She noted that the details of such computation still remain unknown to neuroscientists. Nevertheless, Dr. Churchland said, we can begin distinguishing in-control individuals from those who are out-of-control by analysis of their "parameter spaces"—the combinations of different parameter values that position the organism in one state or another.

PATRICIA SMITH CHURCHLAND: When we think about "the self" from a neurobiological perspective, it appears that we really should be talking not about one particular thing—some single entity that is the self—but rather about a multidimensional affair. The "self" is a set of capacities that involve not only representation of the body itself but also representation of internal aspects of the brain—the brain's mental life.

This set encompasses such disparate things as our autobiography, what we currently feel about our body configuration, where we are in space and time, where we rate in the social order, and the status of our relations to other humans and nonhumans.

It has been argued, particularly by Hanna and Antonio Damasio, that the platform for an animal's most basic self-representational capacities is in the brain stem. This circuitry handles the fundamental problem of coordinating one's needs with one's internal milieu so that the body can move appropriately—to feed, flee, fight, or reproduce. Movement decisions must be elaborated so that you aren't feeding when you should be fleeing and to ensure that you do not try to do incompatible things.

Also within that basic brain-stem platform—in mammals at least, but probably in birds and other vertebrates as well—is a capacity to do motor planning. Organisms need to do some of the figuring out of how to solve a particular motor problem offline—to conduct much of the trial-and-error business in a safe environment—namely, within the brain itself. This sort of motor planning appears to involve the development of an inner model. The work of David Wolpert and also of Rick Grush suggests that increasingly fancy inner modeling gives us the basis for imagining what can happen not only in a complex motor situation but in a social one as well.

Here's a brief sketch of what an inner model might look like: A goal state will be specified. It might be something as simple as "Can I reach that plum?" or something slightly more complicated, like "How do I hide from that predator?" The inner model basically

Dr. Patricia Churchland, University of California, San Diego.

proposes a quick and dirty suggestion about how best to achieve the goal. It then sends a signal to the "emulator," which essentially says, "If you do that, these will be the consequences." This information is then cycled back to the inner model, which can upgrade the initial solution: "Well then, let's make a modification." The new plan will go to the emulator, which may then suggest consequences that are more self-serving. Ultimately—after this kind of back-and-forth iteration converges on a satisfactory solution—a signal is sent to the body and there is a behavioral outcome as the plan is executed.

In brief, the wiring yields self-simulation with respect to the things in the world. But some of those things—at least for those of us who are social creatures—will entail the simulation of other *selves.* What will that organism do if I display anger? What will it do if I chase it? If I try to eat it? The simulation within this initially rather simple emulator structure can get very elaborate, as wiring permits.

Emulators may also, of course, involve self-control, so that the organism can make a decision that best serves its interests. Evaluation of self-interest will take into account, of course, not only immediate

Dr. Churchland: Neuroscience still does not know the neural basis of morality.

needs but also the long-term consequences of each considered action. It is perhaps not too surprising that one can conceive of the development of *conscience* within this very general structure of the emulator or inner model. In order for an animal to come to a fast decision about whether to do one thing or to reject that option and try to formulate another, relevant perception and relevant memory have to be fed into the emulator, and relevant computation must ensue. And all that seems to have a lot to do with, and to be greatly guided and optimized by, the presence of feelings generated in response to the inner modeling of an option. To a first approximation, what we call conscience is the negative feeling evoked by emulation of a social action.

How exactly any of this is done remains puzzling. In particular, we have little idea at this stage of the exact nature of the relevant computation. For example, instead of running through all *possible* options, which the organism clearly does not do, the brain manages to confront and deliberate on only the *pertinent* options. How "relevance computation" works is not well understood.

My sense is that the details of decision making, of choice, of acquisition of character and temperament, and of development of such things as moral character are going to elude us until we have made more progress on certain fundamentals of neuroscience—namely, the dynamic properties of neural networks. At present, there is an enormous gap between what we know about how *neurons* work, and what we know about *networks* of neurons.

Still, let me say just a little bit more about neuroconscience, though necessarily at a very general level. Whatever else it is, if the neuroconscience is connected to the emulator, it has to somehow also be connected in a very profound way to the reward-and-punishment system. It must involve simulation of injunctions and warnings, in the way that Socrates said that he heard a little voice telling him not to do immoral things. It must also involve what's sometimes called the theory of mind—the recognition of others as having beliefs, feelings, and desires. In other words, it involves the manipulation and use of those social emotions that Antonio Damasio talked about. Let me turn now to the related issue of making rational or self-interested choices.

Ultimately, we'd like to have some general understanding of the neural differences between someone who is operating with what we might loosely call free choice and someone who is not.

Ultimately, we'd like to have some general understanding of the neural difference between someone who is operating with what we might loosely call free choice and someone who is not. Another way of putting this is that we want to understand the neural difference between someone who, roughly speaking, is *in control* and someone who, also roughly speaking, is *not in control*. We are beginning to understand some of the relevant parameters: levels of serotonin, levels of dopamine, hormones, the wiring between the amygdala and ventromedial frontal structures, leptin concentrations in the blood. For example, low levels of serotonin are associated with reckless behavior in monkeys. Leptin-receptor deficits correlate with obesity. Ventromedial frontal damage correlates with failure to evaluate consequences.

When we come to better understand these parameters and their role in rational choice, even if it's only at a general level, we can begin to think about the in-control versus out-of-control distinction in terms of a "parameter space." That is, each of the parameters (whatever they turn out to be) constitutes an axis in that space. This means we can start characterizing the volume within that parameter space wherein live the in-control brains. The boundary is probably not well defined.

There are undoubtedly many different ways of being in control; different combinations of parameter values will work equally well. Some people may manage to be in control when their serotonin levels are here and their dopamine levels are there, even while they have a rather tenuous connection between the amygdala and ventromedial frontal structures. Others might have different profiles but still be within the in-control volume of the parameter space. The relationships between the parameters are also a target for research.

In the long run, I suspect that we will be able to find general and, ultimately, highly detailed ways of distinguishing between the in-control brain and the out-of-control brain. Notice that in all instances the behavior is *caused* by brain events. At the level of the neuron and the neural network, the brain is a causal machine. Nevertheless, the fact of causality in the brain does *not* imply that there is no responsibility. The determination of responsibility within the criminal justice system depends on many factors, including efficacy of punishment, public safety, and the social importance of retribution.

Neuroethics: Reductionism, Emergence, and Decision-Making Capacities

SUMMARY: "What I've tried to do is present several pieces of my approach to some central neuroethical issues," Dr. Schaffner said in summarizing his presentation. "I dismissed sweeping reductionism and said that creeping reductionism is what neuroscientists do. I rejected sweeping determinism but accepted the prospect of creeping determinism to address moral-choice problems. I proposed we might look at those moral choices in a practical way by generalizing the notion of 'excusing' or 'invalidating' conditions, following H.L.A. Hart. And I proposed that the model of the MacArthur Competence Assessment Tool (MacCAT) might be one way to begin thinking about these things in a focused manner. I also urged that we look at ways, perhaps guided by neuroscience, that emotional capacity might be brought into these MacCAT instruments."

Dr. Kenneth Schaffner, George Washington University.

KENNETH F. SCHAFFNER: In his book *Descartes' Error*, Antonio Damasio says, "The fact that acting according to an ethical principle requires the participation of simple circuitry in the brain core does not cheapen the ethical principle. The edifice of ethics does not collapse, morality is not threatened, and in a normal individual the will remains the will."

But because simple brain circuits don't prima facie generate moral decisions, this comment suggests that we at least ought to take a look at the philosophically perennial issues of free will and determinism, as well as reductionism and emergence.

Other authors are not as sanguine as Dr. Damasio. For example, Daniel C. Dennett, in his book *Consciousness Explained,* says that "if the concept of consciousness were to fall to science, what would happen to our sense of moral agency and free will? If conscious experience were reduced somehow to mere matter and motion, what would happen to our appreciation of love and pain, and dreams and joy? If conscious human beings were just animated material objects, how could anything we do to them be right and wrong? These are among the fears that fuel the resistance and distract the concentration of those who are confronted with attempts to explain consciousness."

Now, what Dennett suggests is that we *not* be distracted. And what I'm going to try to do is to clear away some distractions by making some distinctions. That's what philosophers do; they make distinctions.

One distinction is between two kinds of reductionism. The first is what I call sweeping reductionism, where we have a sort of theory of everything and there is *nothing but* those basic elements—for example, a very powerful biological theory that explains all of psychology and psychiatry. The second kind is "creeping reductionism," where bit-by-bit we get fragmentary explanations using interlevel mechanisms. In neuroscience, this might involve calcium ions, dopamine molecules, and neuronal cell activity, among other things.

Sweeping reductionism, I think, is probably nonexistent except as a metaphysical claim. There is some scientific bite in trying to do something like this in terms, say, of reducing thermodynamics to statistical mechanics, but these sweeping reductions actually don't work when you press for details. They tend to fail at the margins. So I don't think that sweeping reductionism really has much in the way of cash value. It's a scientific dream.

I don't think that sweeping reductionism really has much in the way of cash value. It's a scientific dream.

Now, creeping reductionism, which I favor, can be thought of as involving *partial* reductions. Creeping reductionism is what neuroscientists do when they make models and propose mechanisms. Creeping reductions do not typically commit to a nothing-but approach as part of an explanatory process. Rather, they seem to tolerate a kind of pragmatic parallelism, or emergence, working at several levels of aggregation and discourse at once. And creeping reductions are consistent with a coevolutionary approach that works on many levels simultaneously, with cross-fertilization.

Clearing away another distraction requires distinguishing between two kinds of determinism. "Sweeping determinism," regarding a powerful theory we think might be fundamental and universal, states that given a set of initial conditions for any system, all subsequent states of the system are predictable and determined. This is what some Newtonians believed.

And quantum mechanics, though it's indeterministic in the small, is essentially deterministic for bodies, like cells and organisms, that are medium-sized and larger. In the genetics area, where we focus on the presence of powerful genes (alleles) related to disorders and traits, this kind of determinism is called genetic determinism.

But sweeping genetic determinism has so far failed to be the case empirically, and sweeping neuroscientific determinisms are not yet even close. We heard that from the first two speakers. What we do have is "creeping," or partial, reductions to neuroscience, with determinism that may be coming. As mechanisms are elaborated, neuroscientists will get roughly deterministic explanations for some types of behavior in some people. And I say "roughly" because here too there will be some problems at the margins.

Claims of sweeping determinism worry philosophers and the philosophically inclined, but a mechanical determinism of a sweeping sort has never had any legal relevance, so far as I know. Nobody ever brought somebody into court and said that they were mechanically determined by Newton's theory of motion. But a concept that does have legal bite regarding free-will issues is what's called in criminal law "excusing conditions" and in civil law "invalidating conditions," to use some of H.L.A. Hart's terms.

In this view, which I think I favor, attributing free will to an individual is the *default position.* But the presence of excusing conditions, some of which might be subtle and informed by neuroscience along the lines of what we've heard from the first two speakers, would imply *lack* of free will. Excusing conditions, such as loss of muscular control, subjection to gross forms of coercion by threats, and types of mental abnormality, may make an action unintentional. They are believed to render the agent incapable of choice.

Dr. Schaffner: Should there be "excusing" or "invalidating" conditions affecting moral choices?

Regarding determinism from this point of view, Hart writes that "there will be disputes as to what conditions to admit as excuses, and what degree of proof to require." But we have examples in case and statute law to assist this. I think that we have similar ethical principles, cases, models, and analyses in the moral realm to demarcate those things where we are willing to attribute free will operating and those parts where we are not.

One such model we might consider for bioethics—one that incorporates a set of principles bridging the legal, regulatory, and moral realms—is work done on decision-making capacity both in therapy and clinical research. This is a well-developed test instrument known as the MacCAT, which stands for the MacArthur Competence Assessment Tool. It looks at four components relating to informed consent: understanding, appreciation, reasoning, and expression of choice. The MacCAT has several flavors that have been applied to depressed and schizophrenic patients.

We can interpret the MacCAT, within the context of this conference, as a test for the excusing or invalidating conditions that would exclude such patients or, in the conditions' absence, affirm patients' free and informed choice to accept therapy or participate in clinical research. And I think we can generalize further and suggest that similar formal and informal tests govern our notions of moral culpability in all ethical situations.

But a question that arises is, Does the MacCAT in its current form look at all the relevant conditions affecting free and informed choice? One objection is that emotions were left out of the MacCAT. And in response to that criticism, Paul Appelbaum has replied that some kind of capacity for emotion is required. He's aware of Damasio's comments about the need to move in this direction, but he says that Damasio's cases, such as Phineas Gage and the patient Elliot, are rare and that we would need to have more powerful arguments to warrant an emotional-capacity dimension. Also, we'd need reliable ways to measure such capacities.

Does the MacCAT in its current form look at all the relevant conditions affecting free and informed choice?

Other philosophers, too, have strongly favored the incorporation of emotions within an approach to ethical decision making. Aristotle and Hume, and the present-day philosophers Martha Nussbaum and Pat Greenspan, among others, have argued for the important role of emotions in ethical analyses. I would suggest that one thing neuroethics might well do is to take Damasio's points seriously, amplify them, and try to relate the neuroscience perspective to Appelbaum's analysis. This would involve simultaneous work at several levels, going back and forth in a coevolutionary way.

Gaging Ethics

SUMMARY: Dr. Moreno maintained that the assumption of self-determination is critical to our notions of bioethics. But is it a valid assumption, given that in many situations the exercise of free will or informed consent may not be reliable or even possible? He pointed out that bioethics has a particularly pragmatic and democratic American flavor—that Americans' beliefs in learning from experience and consciously shaping their own will depend on our notions of individual understanding and capacity for choice. And although many grave issues in bioethics are shot through with doubts about self-determination, it is indeed possible perhaps only because we will it to be so. Procedural values, rather than substantive values, are the enablers. Patients can handle the truth if they're told the truth.

JONATHAN D. MORENO: Modern bioethics, which emphasizes patient or subject autonomy and the doctrine of informed consent, appears to have placed a bet. It is that self-determination—a person's conscious expression of his or her own moral will—is an essential part of physician-patient relations and of health care decision making in general. But as empirical evidence increasingly suggests that even relatively healthy people may have impaired decision-making capacity in some cases, bioethics might have made a rather dangerous wager.

Does self-determination work? Who chooses for those who cannot? Is informed consent itself a myth? These questions, which we haven't really worried about very much in bioethics, could precipitate an existential crisis even as I speak—though you'll be glad to know my talk has a happy ending.

Bioethics as it's practiced is a quintessentially American phenomenon. This doesn't mean that bioethics can't be done in other countries but that certain of its elements—themes developed by philosophers such as Ralph Waldo Emerson, William James, and John Dewey—are simply very American. To paraphrase one of Jack Nicholson's movie lines, the presumption in bioethics has been that we *can* handle the truth.

I argue that Americans have placed particular confidence in the faculty of intelligent decision making and in an individual's capacity for free will. (Contrast this belief with what some non-Americans have thought, in particular a nineteenth-century French physician named Thouvenal, who was perhaps the ultimate physician-paternalist: "Who is better qualified to play [the role of deciding how a patient should live] than the physician, who has made a profound study of [the patient's] physical and moral nature?")

Dr. Moreno: Is self determination a valid assumption?

But Americans have from the very beginning had a sense that things could be different in physician-patient relations, and that patients could be self-determining—that they could have a certain moral will of their own. Benjamin Rush, for example, said of medicine during the American Revolution that "the people rule here in medicine as in government" and—a little lesson for our contemporaries who sit on ethics committees and fill out informed-consent forms—that ". . . truth is simple upon all subjects. . . Strip our profession of everything that looks like mystery and imposture, and clothe medical knowledge in a dress so simple and intelligible that it may become obvious . . . to the meanest capacities."

William James worried deeply about free will and determinism. He was obsessed by the problem. And he wasn't sure, although he really wanted to believe that Emerson was right about this, that we could indeed be self-reliant and self-determining. He wrote a wonderful essay, "The Dilemma of Determinism," in which he simply took the position that we make an existential choice to believe in free will as our first act of free will.

When he later wrote *The Principles of Psychology*—a book we should acknowledge at any neuroscience meeting as perhaps the field's first systematic work—James articulated what is still, I think, a fundamental principle of neuroscience: When we think about habits and moral education, it is essential to think in terms of the plasticity of the organic materials of which our nervous system is composed.

This is such an important notion for us. Neural tissues, after all, are plastic, manipulable, and manageable; they are able to yield to certain influences but will not yield all at once. And the idea had tremendous social consequences for James. For example, he writes: "The great thing then, in all our education, *is to make our nervous system our ally instead of our enemy*" and that "habit is the enormous flywheel of society, its most precious conservative agent." He is confident, despite the occasional calcification of our neural systems, that we in fact learn by capitalizing on the acquisitions of experience.

Dr. Moreno: Can the sick or dying give valid consent?

Dewey, of course, picked up on the importance of the role of habit in a moral education. In books like *Human Nature and Conduct* and *The Quest for Certainty,* he elaborates on the importance of habit as a way of shaping our will and capturing our freedom.

Now, the assumption that patients can handle the truth—that is to say, tolerate and act on it—is a relatively recent phenomenon. When I teach my medical students, I like to show them two contrasting surveys on what cancer patients have been told about the nature of their disease. In 1961, when hundreds of internists were asked if they ever used the word cancer with their cancer patients, 90 percent said never or rarely. As the *Journal of the American Medical Association* characterized that survey, "Euphemisms are the general rule."

Eighteen years later, when the same question was

asked in another survey, 98 percent of the respondents said that they always or usually use the word *cancer* with their cancer patients. Now, what caused this difference? Obviously, some major social and cultural changes—including medical changes—occurred in the 1960s and 1970s. But the point I want to make here is that as modern bioethics emerged, it manifested the popular cultural notion that patients can handle the truth if they're told the truth, that they can be self-determining, that they can make appropriate judgments for themselves.

Yet the most grave issues in bioethics are shot through with doubts about self-determination. Can parents, for example, give competent consent for sick children? Or are they simply too emotionally tied to what their children are going through to be able to do so? Who may permit experimental and perhaps nonbeneficial treatment for incapacitated adults, or for those in developing countries who don't share our presumptions about self-determination and autonomy? Who speaks for the fetus or embryo? Can the sick or dying give valid consent?

In fact, Eric Cassell has suggested that there's a real question whether any sick person can ever give valid consent to anything. Just being ill, and knowing that one is ill, imposes inherent limitations on the ability to make judgments. So does bioethics rest on a mistake? Can the central dogma of bioethics—self-determination—be salvaged?

Well, I think there is a way to salvation—the pragmatic-naturalist way advocated by James and Dewey, and to some extent Emerson. Self-determination is possible to a sufficient degree perhaps only because we *will* it to be so—because it corresponds to the subjective experience, the phenomenology, that we have about being deciders. The index of deciding for others is not going to be a substantive one of "What's the right decision for this person?" but rather "How have we gone about making a judgment on behalf of this person?" So it is procedural, rather than substantive, values that will rule.

Does bioethics rest on a mistake? Can the central dogma of bioethics—self-determination—be salvaged?

Finally, informed consent, even if it's only a ritual that we go through in some instances, is a very important ritual. It expresses what the National Commission for the Protection of Human Subjects calls respect for persons. It may have to be taken in what Richard Rorty would have called an ironic spirit. But this is a particular kind of ironic spirit, one in which we take the principle very seriously even though we know that it may not always work out in practice the way we'd like.

Question and Answer

FROM THE FLOOR (unidentified speaker): Are all emotions "good" emotions?

ANTONIO R. DAMASIO: Actually, there are many emotions that do not seem all that "adaptive"—in fact, they might lead to rather counterproductive results from the point of view of ethics. I think that emotions *began* as highly adaptive—as highly positive problem solvers—but depending on the circumstances, they may not actually produce a positive result.

Just take the example of fear. There is no question that fear is one of the most powerful, highly adaptive emotions throughout the history of living organisms. And yet, in many circumstances, for many of us, fear is misapplied. I'm not even talking about the extreme situation of the maladaptations we call phobias. But the fact is that very often we end up engaging fear when it is of no value to us whatsoever. This is also true, across the board, for the so-called social emotions.

There is no question that fear is one of the most powerful, highly adaptive emotions throughout the history of living organisms.

FROM THE FLOOR (unidentified speaker): The emotions you're discussing don't even need to be "social" to affect social behavior, right?

DAMASIO: I think that's a very important point. Disgust—which is not, to begin with, a social emotion—is a good example. It's a wonderfully prepared mechanism to help you reject bad proteins, for instance. And yet we all talk about "disgust" in reference to a social event. We have transported that particular package of biological reactions to the social realm, and created a metaphor.

FROM THE FLOOR (unidentified speaker): Is a social emotion, or any emotion, a necessary prerequisite to appropriate social behavior?

DAMASIO: The answer to your question is that in

many circumstances the two can be dissociated. So you can have individuals who do not produce a so-called social emotion and yet can evaluate a social situation in incredibly rich detail. But this varies with the circumstances. When the social-emotion system goes down, the likelihood of triggering an automatic analysis of the social situation goes down with it.

Drs. Moreno, Schaffner, Churchland and Damasio.

You can force a nonautomatic deliberate analysis. For instance, we've done experiments with individuals who we know fail to decide correctly in certain kinds of social scenarios—in real time and in real life—and we ask them to consider those scenarios in a prepackaged, verbally presented form. We then ask them to analyze the scenario and tell us what they would do. And what is interesting is that under laboratory circumstances, the adult patients *can* often make the right analysis and do very well.

So the failures of those patients really have to do with the engaging of the emotion-triggered "automatic system" that pervades many social behaviors. There's a difference between the automated and the deliberate, and these people just cannot force themselves into making the analysis in real situations.

WALTER GLANNON (McGill University): Much of the discussion this morning was focused on what philosophers would call the negative sense of free will. That is, we assume people have free will when there's no neurological dysfunction and they can voluntarily perform a bodily movement. But some worry about the positive side of free will, when it's not enough to just do an action but we have to have second-order desires as well—for example, that the action will affect our praiseworthiness for the positive things we do. That is, we want to think of ourselves as authors and originators of our actions, where those actions follow from mental states. But if there's a neurobiological basis to all that, a lot of philosophers are concerned. So what would you say to them to allay their fears?

DAMASIO: If I understand your question, you're worried that as we understand more and more of the neurobiological mechanisms that result in all forms of behavior and all forms of mind processes, we somehow get deprived of the authorship of those actions and of those thoughts.

Well, my first reaction is: if that comes to pass, so be it. But I am not convinced that it is going to be quite that way. Our last speaker referred to the idea—attributed to James, Dewey, and Emerson—that you can somehow force yourself into an illusion that you are the author of those thoughts. And you can preserve that illusory framework for the practical purpose of being a human being in a society.

And it may be that we are already doing a lot of this, given all those humiliations we have endured for the past four centuries in terms of our notion of what we are and where we stand in the whole scheme of things.

GLANNON: Would we be deluded into thinking that we were praiseworthy individuals if we simply thought that way?

PATRICIA S. CHURCHLAND: Actually, I think it's quite reasonable to expect that nothing much will change with respect to taking pride in one's achievements, or being ashamed of one's failures, or whatever it happens to be. You don't even need to do what Tony was suggesting—namely, make a conscious decision that you're going to act as if you are a person. The brain has already done that for you; it's part of the brain's "user illusion," so to speak. So I think you can just carry on more or less in the usual fashion.

But I think there *will* be times when self-understanding about how it is that you are the person you are, and the factors that went into making you that person, will deepen your understanding of yourself and allow you a degree of freedom that you wouldn't otherwise have. So I see the developments in neuroscience as actually enhancing our sense of self, rather than diminishing it or taking power away from us.

How can we more systematically and reliably recognize and address the unconscious bias inherent in our science?

STEPHANIE J. BIRD (MIT): One of the things that really concerns me as a neuroscientist is that our internalized notions frame so much of what we think are the questions to ask and what we recognize as the answers. So I'm wondering if you all have some thoughts about how we can frame questions that are less wedded to those notions. Specifically, how can we more systematically and reliably recognize and address the unconscious bias inherent in our science?

CHURCHLAND: What the history of science suggests is that you start with what you've got. You try to identify the problem as best you can, and then you bootstrap your way along, and you reconfigure your question.

Consider seventeenth-century physician William Harvey. He investigates the heart to try to understand exactly where and how the "animal spirits" are concocted. That's what motivates his work on the heart. He comes to realize as he's working on the heart and trying to understand how it is that the animal spirits are concocted there that the heart is just a pump. To his surprise, Harvey recognizes that not only are the animal spirits not concocted there but in all probability "animal spirits" do not exist. So he comes out at the end of his exploration having completely reconfigured the question. Now, given the falsity of the "animal spirits" paradigm, Harvey wants to know the difference between arterial and venous blood, what exactly the lungs do to blood, and so on.

Those sorts of surprising conceptual revolutions happen time and time again, and we can see them in neuroscience as well. For example, we don't think of memory, as many people did roughly one hundred years ago, as a kind of uniform, big bank. Now we know that there are many, many different kinds of plasticity, and that plasticity varies over time, and that there are different systems, and so on. The concept of memory has changed as a result of research in neuroscience and psychology.

So the general point is that there isn't any easy way to discover the nature of reality behind appearances, except to slog away at it and be prepared to change your mind if that's how the facts go.

The concept of memory has changed as a result of research in neuroscience and psychology.

DAMASIO: I would second that. We are at the point of marching forward with more and more findings. Of necessity, our hypotheses will be reconfigured as we continue. And they'll probably be reconfigured many times.

For example, we are now ready to accept that emotions are important in our ethical behavior and in our social behavior in general. There has been a dramatic change over the last twenty years, and especially the last ten. Previously, there was an almost complete exclusion of anything having to do with emotion—or feeling, or consciousness—from the realm of neuroscience. They simply could not sit at the table.

This was also true in modern philosophy until recently. In fact, Ronald de Sousa exemplifies an exception: he was talking about the value of emotions in ways in which many contemporary philosophers were not. We are going through a very big transformation, I think, both in philosophy and in neuroscience.

KENNETH F. SCHAFFNER: But I think one has to be careful about how much diversity there is or isn't.

My sense is that there's still a fairly strong cognitivist bias in analyses in medicine and such. The informed-consent approach that Jonathan was talking about represents that cognitive bias, which doesn't really pick up on the emotional dimensions. By *diversity* I mean the appreciation of the role of the emotions in addition to cognitive factors.

I had mentioned that Appelbaum is resistant to incorporating them into his instruments because he doesn't know how to test for them. So we might all acknowledge their importance, but first we have to grasp them and understand the right way to proceed. And then not overreact. I had a sense in some of the discussions that people were going to the other extreme and saying, well, *everything's* emotion.

ANITA SILVERS (San Francisco State University): I'd like to go back to Pat Churchland's very important notion that having a neuroconscience means recognizing others as having feelings and minds. I'm not sure whether that's a necessary condition of having a neuroconscience, or just a central notion, a central ability of a neuroconscience.

People with Asperger's syndrome do not recognize others as having feelings and minds...Do they have neuroconsciences?

I'm thinking about two different pathologies in which individuals are not able to do this. But the phenomenology seems quite different in the two cases. People with Asperger's syndrome do not recognize others as having feelings and minds. They are, however, able to train themselves to follow certain kinds of social rules. They do it on a basis, I think, quite different from individuals who are empathetic. Do they have neuroconsciences? Individuals with traumatic brain injuries, however, at least some of them, are also unable to recognize others as having feelings and minds. But their phenomenology seems to be very different—that is, their descriptions of their awareness of other people are very different from the first case. And they have much less control over how they respond than in the first case.

So I wonder if you could talk a little bit about this notion of recognizing others and provide some more detail about what's involved in that.

CHURCHLAND: To a first approximation, someone lacking the capacity to recognize others as conspecifics and as having similar pains and pleasure may still be trained to observe the social norms. This just reminds us of the importance of reinforcement—that is, the value of reward and punishment in learning. And it must play an absolutely crucial role in the development of conscience.

I sometimes worry that we approach these questions with a messed-up paradigm. For example, we make a distinction between the emotions on the one hand and reason on the other. We may suppose they are two completely different categories: the emotions are handled by one part of the brain; and reason, a mechanism that's thought to act independently, is handled by another. However, in neural reality, the two are probably part of a continuum. So what we call the emotions may simply lie at one end of the

What we call the emotions may simply lie at one end of the spectrum, with what we call pure reason...sitting at the other end.

spectrum, with what we call pure reason—perhaps, when we're doing something like proving a theorem or trying to be tremendously objective about our data—sitting at the other end. They may not be two completely different classes of brain function at all. We'll know better as the data from basic neuroscience really start coming in.

The traditional bifurcation of emotion and reason is of course extremely useful on a day-to-day basis, especially in training children. It's convenient to be able to say, "Don't let your emotions run away with you;" "Don't panic;" "Use your reason;" "Think it out." Nevertheless, my hunch is that this distinction is like the medieval distinction between sublunary physics and superlunary physics: the idea that the laws of physics that applied from the moon down to the "center of the universe" (namely, Earth), and different laws that governed the properties of the perfect things beyond the Moon. It turned out, after Kepler, Copernicus, and Newton, that there was no distinction of any interesting kind in the physics of the sublunary realm and the superlunary realm. I suspect there is no interesting psychoneural distinction between emotions and reason.

Consequently I have this feeling that a big, big reconception on the nature of decision and choice is in the cards for neuroscience and psychology. It scares the hell out of me, but that's what I think is coming.

DAMASIO: I entirely agree. I used to say, in fact, that emotion was in the "loop of reason." In spite of the fact that under certain circumstances some emotions can be counterproductive and nonadaptive, emotions are packages of reactions that have been well preserved in evolution. Emotions have an inherent rationality, at least within a very broad number of contexts, and are quite valuable in particular environmental circumstances for the organisms that exhibit them.

So emotions are the beginning of reasoning. But of course they don't operate effectively when the environment and problems get very complex—when the solution of the problems requires a lot of creativity and intelligence on the part of the operator. It is in those circumstances that cognition gets in.

Given the brain's high degree of redundancy, you have the possibility of compensating for degeneracy in certain areas.

I have a sense that evolution took another step for decisions requiring what we call reason. I think, in fact, that it added on elements to the process. But emotion continued to be important, for example, in narrowing down the decision-making space—letting us get very quickly to some choice that is highly biased by what we have learned in the past about emotional reactions in certain circumstances. So I really see emotion and reason as continuous processes.

I wanted to make one other point. When we talk about neurological conditions, whether it's Asperger's syndrome or the result of traumatic head injury, we fall into the trap of thinking that they are monolithic. We talk as if certain things that people with those conditions typically cannot do—for example, sense feelings in others—are permanent impairments. And that's just not so. The circumstances—the context—

can change the appearance of a symptom or suspend it. Also, given the brain's high degree of redundancy, you have the possibility of compensating for the damage with the work of other brain systems. These other systems allow you to do certain tasks though maybe not the same way or as well.

JONATHAN MORENO: In the unaccustomed role of historian of philosophy, I think it's worth putting on the table another tradition that did a very good job of not trading too much on the distinction between rationality and the emotions—namely, the Scottish "moral sense" philosophers Shaftesbury, Hume, and Hutcheson. Much of what they had to say about the even distribution of outliers from one culture to another, in my understanding at least, is now being confirmed by social psychologists and criminologists, who have shown that most crime in any society is committed by a small minority of repeat offenders. This suggests that for most of us the emotions and rationality work together rather well.

FROM THE FLOOR (unidentified speaker): One of the things about being an effective learner is that once you've learned something you don't want to have to keep relearning it over and over again. We take in new information, we respond, and then many things become automatic. They seem to be beyond our consciousness, beyond our free will—which appears to exist in a narrow window of information processing that is our consciousness. If that's the case, then free will is limited to our exposure to novel experience, novel concepts, novel questions—things we haven't addressed before. I'm wondering if you could comment.

CHURCHLAND: I don't think that the only time I'm exercising free will is when I'm agonizing, but for all my "in control" behavior. I'm exercising free will right now when my words are coming out of my mouth, and I don't have any chance to deliberate. If I do deliberate upon my next utterance, then of course I get tongue-tied and I make a real hash of it. The brain is a dynamical system, and it's extremely difficult to predict moment by moment what somebody is thinking, what they're going to choose, what their decisions will be. So predictability and causality are not the same. Some caused events are not predictable. The way I think about the "domain of being in control" is that it's a region in which somehow or other, given that the brain is a dynamical system, it has found a certain kind of stability relative to the parameters we know are important.

SCHAFFNER: It doesn't seem to me that your having conditioned yourself into a set of habits, and these habits' inertia, is necessarily opposed to free will. It may be that you self-identify with those habits and that's what you want to do. And in fact if those habits go awry, you say, "Gee, I wish I could get back into that way of doing things." And that's a second-order question, not necessarily related only to the inertia of habits.

CHURCHLAND: That's a good point.

JODI HALPERN (University of California, Berkeley): I've worked for many years on a book about empathy in the patient-physician relationship. The central question driving my book is: What role does one's own emotion play in learning more about another person's emotion? In other words, not just reading someone's face and saying that the person is angry, or that the person is afraid, which we know very well can be done from a detached state of mind. But the idea that there's something very innovative within emotional experience that allows an openness to seeing things in new ways raises a philosophical problem that I don't know an answer to, and I wonder if the philosophers on the panel may have ideas about it: Could a co-resonant emotional state guide cognition toward information about another person that would otherwise be unattainable?

What role does one's own emotion play in learning more about another person's emotion?

DAMASIO: I just don't see the two possibilities you mention as competing at all. I see them as perfectly compatible and, in fact, in sequence. Here's why: I make a distinction between emotion and emotional experience, which I call feeling. I think that there is an ordered way in which these phenomena have come about, in terms of the evolution of biology, and the individual's own development.

I don't think that when we have an emotion the process stops there. The process continues with feeling. And feeling in itself has repercussions in terms of the way we operate. And of course, when you talk about empathy, you talk not about an emotion but a feeling–the emotion is sympathy. There are, for example, certain thoughts that are co-evoked along with the feeling of empathy. And there will be elaborations, thinking elaborations, following on the heels of that feeling. And all of those will have an enormous impact on the way you govern the subsequent behavior. So I don't see a conflict at all.

I don't think that when we have an emotion and a subsequent feeling, the process stops at emotion.

SCHAFFNER: Could you give us an example of where these emotions have this kind of innovative role?

HALPERN: A patient I've worked with has Guillain-Barré syndrome, and as he was speaking to me it became clearer and clearer that there was something just very shaming in the entire experience. He was paralyzed. He couldn't move his body. This was a very powerful man, whose wife and grown daughters were seeing him in this state, unable to move anything and constantly being instrumented in the hospital room.

Everyone saw him as just too tired or too weak to be communicated with. I too began speaking to him in a quiet voice, which was only making him quieter and quieter. But I too started feeling incredibly embarrassed and ashamed—a resonant kind of shame. And as I was feeling this shame, I started to think about his words in a different way, and I started to realize that I'd better change my tone. And it pushed me to start talking to him in a much more assertive way, and ask: What's really bothering you about the way we're treating you here? What's bothering you about the experience? This made him much more comfortable, and able to speak to me in a very angry way about all the doctors, nurses, and students interrupting him. And he just changed completely in his ability to communicate with me.

But I think that if I hadn't felt this shame, which apparently many other people did not feel or weren't conscious of, it wouldn't have redirected me. So that's an example.

SCHAFFNER: It's helpful. I think what we need to do in this area is to focus on good prototypes and examples so that we can begin to build the right kind of language and make the right distinctions.

MORENO: What Jodi's bringing to our attention is a debate within clinical bioethics between people who support what is sometimes called the "care perspective" and those who are more interested in doing clinical ethical reasoning through casuistry or through the vaunted principles of bioethics with self-determination as the trump.

COLIN BLAKEMORE (University of Oxford): I wonder if the panel members could give their views on the distinction that Ken drew between sweeping and creeping determinism. Personally, I think I'd rather be a sweep than a creep. Sweeping determinism has actually been rather useful as a tool. In our explanations of the rest of the world, it underpins the scientific method; science has tackled many complex problems before, moving from ignorance to deterministic explanation. As thinking beings, we all realize the worries and the social consequences that would spring from a deterministic account of human behavior. But really, what are the alternatives?

SCHAFFNER: When I said that I didn't want to accept sweeping determinism, I called it a metaphysical thesis. And I don't dis metaphysics. I do metaphysics a lot. I'm a pragmatist, so it plays an important role in my philosophical approach. I just didn't want people to think we had that kind of sweeping reductionism on the horizon in the area of the neurosciences. That was one point.

The other point I made was that it was groundwork for methodological approaches. People do believe that there's some kind of a basic level, and they try to get to it. On the other hand, they don't ever really get there. There are always these patchy results, but they're guided by a background monism of some sort. A metaphysical claim doesn't have much power in the laboratory except as a methodological rule, but it's a prime mover for practical advances, even breakthroughs.

CHURCHLAND: I take a different view from Ken on this. I think that what you want to do in science is look for reductions. To get explanations and understanding you really do want to go from the macro levels to the micro levels, and then from each micro level to the next micro level. Determinism (events are caused) is a guiding methodological principle, and I don't really think of it as metaphysical but simply as an empirical hypothesis. However, we know that at the subatomic level there's reason to think that determinism may not obtain. But beyond that it does obtain.

I don't think that in neuroscience we're going to be able to explain something like temperament or storage of a memory in terms of something more basic than neurons. So I guess I am a sweeper in Schaffner-speak. It looks to me like a very reasonable empirical hypothesis that that's the way the world works. It might turn out that there are odd forces here and there, or that some of your explanatory hopes will be thwarted because the mathematics is too hard. But I don't place much significance on the so-called patchiness of the sciences. I think that philosophers, by pointing out that this part isn't complete and that part isn't complete, have tremendously overblown the difficulties of getting reductions. I always want to say: Well, we'll get there. We'll just keep hacking away at it on the assumption, reasonable enough given the history of science, that eventually we'll get the answer or to a good approximation.

I don't think that in neuroscience we're going to be able to explain something like temperament or storage of a memory in terms of something more basic than neurons.

FROM THE FLOOR (unidentified speaker): Dr. Damasio described very nicely the patients who had early injuries in the orbitofrontal cortex and then exhibited antisocial behaviors through their lives. Do you think those patients are, in the sense of Dr. Churchland's remarks, out of control? Do they lack free will for their moral choices? Second, if one were to do a good functional neuroimaging study and show variability among intact brains in the engagement of these same systems during moral decision making, would one then have a measure of variability among humans in their capacity for moral choice?

DAMASIO: Those are both good questions. By the way, over a period of some two and a half years after publication of those first two cases, a couple of dozen other cases have surfaced. And that's exactly what we wanted when we published those two cases, because we were certain there would be others out there.

In the patients I have seen, I'd say that they can be described as out of control. Again, they're not out of control in every circumstance, in every context, but they are out of control a good part of the time regarding what we would describe as ethical behavior.

If I understood your second question correctly, and combining it with Pat's suggestion of a parameter space, I do think it will be possible to find some kinds of parameter combinations that will be reflected, for example, in imaging patterns. I'm just a little bit worried about jumping too quickly from one to the other. There are many pitfalls in the ways we set up experiments and analyze the data from any functional imaging experiment, even if you're doing the best you possibly can.

Still, I entirely agree with Pat that we do what we can. We're making progress. But don't expect that we're going to capture the entire reality and the entire realm of possibilities right away.

HOWARD FIELDS (University of California, San Francisco): I just wanted to add something to the point that was raised by Colin and addressed by Patricia. It is that reductionism and determinism are different. The key thing about the nervous system is that you cannot find the explanations for behavior by looking only within the brain. Some of the causes will be intrinsic, others extrinsic. So, as an example of an intrinsic cause, if I lower the glucose concentration in the hypothalamus, I'll elicit feeding behavior in animals. But particular feeding behavior will be determined by the presence of food and the location of food (extrinsic causes). At the human level, what I eat depends on what's available, and how I eat depends on who I'm eating with.

At the human level, what I eat depends on what's available, and how I eat depends on who I'm eating with.

DAMASIO: I think that's a very good point.

CHURCHLAND: Yes, that seems entirely reasonable. And ultimately, of course, that should be part of the story. But getting all those details is going to have to wait.

DAMASIO: That's why I was saying at the beginning that one should not be worried that we are looking for mechanisms *related* to ethics in the brain. That's simply a very good place to start; it doesn't mean that what we call ethics is generated by brain phenomena alone. It's obvious that the collective interactions of individuals—and in the case of humans, the interactions within a culture—have shaped those phenomena to begin with, and are shaping them still.

ALBERT R. JONSEN: As the chairman of this session, I am going to authorize myself to make the final comment. My intent is to put this entire discussion in the wider context of intellectual history. Its extreme generality verges on parody, perhaps, and invites challenges from informed persons, but I believe it contains some truth. Until recently, the primary concern about determinism in the history of thought was about providential determinism in the major philosophical and religious traditions. By "providential determinism," I mean the way in which some universal divine reality determines the fate and behavior of human beings. Hindu karma is clearly a determinism. Stoicism taught a cosmic determinism. Early Christian theologians, affected by Stoicism, debated how an omnipotent deity and personal sin were compatible ideas. The theologians of the Reformation argued about predestination. It is only since the Renaissance that the determinism debate has turned to what might be called embodied rationality—that is, how do the apparently "spiritual" functions of rational thought and free choice fit within a material body and world? Our discussion today is a version of that post-Renaissance debate. Yet our discussion takes a new, perhaps postmodern turn. The older cosmic question could not be falsified by any empirical evidence whatsoever. Theological claims about a supreme power of the cosmos or of an almighty deity cannot be proven or disproven by data. However, this post-Renaissance and postmodern question appears to depend radically on evidence. There are many ways to

We may be on the verge of a rather new way of thinking about an old problem.

support and to counter the statements made this morning by appealing to empirical data. The particularly postmodern form of the question arises from the vast and increasing torrent of data about mind-body function. We may be on the verge of a rather new way of thinking about an old problem.

Session

II

Brain Science and Social Policy

Barbara A. Koenig, Session Chair
Associate Professor of Medicine and Executive Director, Stanford University Center for Biomedical Ethics

Daniel L. Schacter
Professor and Chair of the Department of Psychology, Harvard University

William J. Winslade
Professor of Philosophy and in Medicine, University of Texas Medical Branch at Galveston

Henry T. Greely
Professor of Law and Director of the Center for Law and the Biosciences, Stanford University

BARBARA A. KOENIG: We heard a lot in the first session this morning about what neuroscience may tell us about human nature. Now we turn to a more concrete discussion of some of the social implications of neuroscience. In particular, studying the brain offers a seductive promise of prediction—the ability to make assessments about people and their motivations, desires, and characteristics. For example, it might be possible to predict the onset of someone's poor cognitive functioning later in life.

Predictions will span a range of other domains as well, including future ill health, potential success in school or employment, violent behavior, or even addiction to drugs. As one of the few social scientists at this meeting—I'm an anthropologist—I'd like to make the important point that whether or not those predictions prove to be scientifically accurate may be less

Dr. Barbara Koenig, Stanford University.

important than our belief in their power.

So this session charts the range of social policy issues that will be affected by our belief in the possibilities of neuroscience to both explain human nature and predict the future. Of special concern will be the drawing of boundaries between the normal and the pathological—a truly important issue in the social policy arena. Our speakers today will focus in particular on law, education, and health care.

We're going to start with Daniel L. Schacter, a psychologist, who works on psychological and biological aspects of human memory. He is a professor of psychology at Harvard and also serves as chair of the Psychology Department there. His most recent book is *The Seven Sins of Memory: How the Mind Forgets and Remembers,* and that is his subject today as well.

The second speaker will be William J. Winslade, who is a lawyer and a psychoanalyst. He is a professor of philosophy in medicine, and he's in several departments—Preventative Health and Community Medicine, and Psychiatry and Behavioral Sciences—at the University of Texas Medical Branch at Galveston, where he is also a member of the Institute for the Medical Humanities. He's coauthor of the book *Clinical Ethics* (with Al Jonsen), and his most recent book is called *Confronting Traumatic Brain Injury,* which will be his topic today.

And last we'll have my colleague Henry Greely—we'll always think of him as *Hank* Greely—who is the Carlsmith Professor of Law and a professor, by courtesy, of genetics at Stanford University. I'm pleased to say we've worked together for almost ten years. Hank directs the Law School's Center for Law and the Biosciences, and he has had a great deal of experience with policy, including service on California's Advisory Committee on Human Cloning.

The Seven Sins of Memory: Implications for Science and Society

SUMMARY: Dr. Schacter elaborated on four of what he calls the seven sins of memory that he said had particular relevance to neuroethics issues. If drugs existed that could reduce transience—the decreased accessibility of memories over time—they would raise equity questions in such settings as schools and workplaces. Absent-mindedness provokes legal issues of just who or what is responsible for a damaging oversight. Similarly, misattribution may cause individuals to be wrongly accused (though early research indicates possible long-term potential for separating true from false memories). And persistence—the retention of undesired memories—could be eased with available drugs. Should those drugs therefore be administered, say, to victim of a violent crime or a disaster relief worker?

DANIEL L. SCHACTER: Memory, of course, is not perfect. There's nothing controversial about that. But it's mostly in relation to the imperfections of memory that interesting questions arise with respect to neuroethics. One way to organize these various imperfections is to think of them in terms of seven fundamental categories—what I refer to as the seven sins of

The Seven Sins of Memory

- *Transience:* decreasing accessibility over time
- *Absent-mindedness:* lapses of attention; forgetting to do things
- *Blocking:* temporary inaccessibility of stored information
- *Misattribution:* attributing memories to incorrect souce; false recognition
- *Suggestibility:* implanted memories
- *Bias:* retrospective distortions produced by current knowledge and beliefs
- *Persistence:* unwanted recollections that people cannot forget

memory. Let's just quickly walk through all seven, and then I'll focus on several of them with respect to their implications for today's proceedings.

The first three sins have to do with different kinds of forgetting. Transience refers to the fact that, all else being equal, memories tend to become decreasingly accessible over time—certainly a prominent type of forgetting. The second "sin," *absentmindedness,* refers to lapses of attention that often result in forgetting to do things, and it's largely distinct from transience. The third of the forgetting "sins" I call *blocking,* which refers to the temporary inaccessibility of stored information—tip-of-the-tongue states and the like.

The next three sins have to do with various kinds of memory distortions—instances in which memory is present, but wrong. *Misattribution* occurs when we link memories to an incorrect source, resulting in a phenomenon called false recognition, which I'll talk about later. *Suggestibility* refers to implanted memories—when, as a result of misleading suggestions, people come to remember events that never happened. *Bias* refers to the fact that we often distort our past in reference to current knowledge and beliefs.

Finally, *persistence* refers to unwanted recollections. This is kind of the flip side of transience, when we want to forget but can't; usually it's the result of some kind of emotional experience.

Of the sins that I think have some of the most interesting potential connections to this conference, let's start with transience. It is often represented by the Ebbinghaus forgetting curve—from the classic experiments of the German psychologist Hermann Ebbinghaus back in 1885—which basically shows that as time passes memory tends to degrade or become less accessible. A pursuit of many researchers now is a drug that would interfere with this transience curve—to keep us "up here" as time passed after learning, rather than coming "down there." We don't have such a memory-enhancing drug yet, but if and when the time comes—and it's more likely when than if—a lot of questions will arise.

If a memory-enhancing drug were able to help children function better in school, •would you want your child taking it?

To begin with, who would take it? Should we all? If a memory-enhancing drug were able to help children function better in school, would you want your child taking it? If the child did not take such a drug, would he or she risk falling behind classmates? What about in business or other job-related settings? For example, to get a job, you know that improved retention of information from a memory enhancer would give you an edge—assuming that the other candidates aren't taking one too.

Dr. Daniel Schacter, Harvard University.

Absentmindedness is a different type of forgetting from transience. It's caused not so much by the gradual fall-off of memory over time but rather by the failure to pay attention when we carry out an act—resulting in such irritating lapses as forgetting where we put our keys or glasses—or by being so concerned with other things, or by simply functioning on "automatic," that we forget to carry out the action we meant to.

For example, a few years ago Yo-Yo Ma left his $2.5 million cello in the trunk of a cab. He had just taken a ten-minute cab ride, got out, paid the driver, walked away, and only a few minutes later realized he had absentmindedly forgotten his cello. Now, this was presumably not a case of transience. If you had said to him as he was leaving the cab, "Yo-Yo, where's your cello?" presumably he would have said, "Oh, it's in the trunk." The information hadn't faded out of memory. Rather, preoccupied with other things, he hadn't given himself a reminder to carry out the action of asking the cab driver to get his cello. Fortunately the New York City police got on the case and he was reunited with the instrument by the end of the day.

But other, darker manifestations of this same basic absentmindedness can raise some interesting ethical questions. Last summer, a woman in Sioux City drove to work in a minivan with her infant daughter, then went to her job while forgetting about the presence of the baby in the back of the van, where she remained the entire day. Unfortunately, the infant died.

Now the question in such cases is, What or who is the responsible agent? When we look at the Ebbinghaus forgetting curve and think about transience, we say, Well, that's just the property of memory; there's really nothing I can do about that. But when we see these absentminded kinds of errors, there's more of a tendency to blame the individual rather than the memory. The responsibility is on you for arranging things so that you don't forget.

In this case the woman was prosecuted criminally, which was surprising to many people. However, the judge concluded that this kind of forgetting, like other kinds of forgetting, should be viewed as a kind of involuntary process—that the woman simply forgot her child and therefore was not to be held criminally negligent. Still, I think there's a larger issue here about who is responsible for various kinds of memory failures, from whatever cause.

Memory distortion—when memory is present but wrong—has already had major legal implications, resulting particularly from misattribution: attributing memories to an incorrect source, resulting in false recognition. A concrete example is the story of the psychologist Donald Thomson, a memory researcher from Australia, who a number of years ago was accused by the victim of a brutal rape as being the perpetrator; the police were led to Thomson on the basis of a detailed and accurate recollection of him provided by the raped woman.

Fortunately, Thomson had an airtight alibi. He could not possibly have committed this rape because at the moment it occurred he was giving a live television interview on, of all things, memory and memory distortion. Ironically, what had happened here was a classic though extreme case of memory misattribution. The woman had actually been watching the interview and then had been raped by an intruder; she merged her memory of the ordeal with that of Thomson's face from the television screen.

But although Thomson himself escaped prosecution, this case raises a more general issue. As we know, much of legal testimony depends on the memory of eyewitnesses, which is not always accurate. What if we had, through our new functional neuroimaging techniques, a way of detecting whether a memory was true or false? Would we want to use it to help adjudicate cases such as Thomson's—when the witness is not consciously lying, he or she is absolutely sure of the (possibly false) memory, and there isn't such an airtight alibi?

What if we had, through our new functional neuroimaging techniques, a way of detecting whether a memory was true or false?

It turns out that there are experimental ways of inducing misattribution errors of this kind, discovered by Deese back in the 1950s and revived by Roediger and McDermott in 1995. Basically what you do is present people with a list of words or terms—such as *candy, sugar, good taste*—that are all related to a word that doesn't appear, such as *sweet.* If I ask them later about an unrelated word, such as *point,* they'll likely recall, accurately, that it wasn't on the list. But if I ask them about a critical lure such as sweet—the word on which all of the word associations converge—I'll get extremely high levels of false alarms on that theme word, accompanied by high confidence.

When people are very confident in their memories, how can you tell the difference between a true memory (in our example, of a word like *taste*) and a false memory (of a word like *sweet*)? Brain-imaging studies we've done over the past few years indicate that by and large the same regions of the brain tend to activate for both—but under certain circumstances some differences are noted. One recent study we did collaboratively with Roberto Cabeza and published last year in the *Proceedings of the National Academy of Sciences* focused on a brain region that's of particular interest to memory research. This is the medial temporal lobe, which includes the hippocampus and the parahippocampal gyrus.

The hippocampus responded pretty much the same to words that subjects were convinced had been on the list, whether this was actually true or not. But we did see a difference in the more posterior part of the medial temporal lobe, the parahippocampal gyrus, which responded nicely to words, like *taste,* that really had been there, but treated the false word *sweet* like a new and unrelated word. This is only one study, of course, but at least now we see that there are differences between true and false memories in the brain, at least under some circumstances.

Can we now use these imaging techniques, in legal and other settings, as a device to tell true from false? Given the current state of research, definitely not. For one thing, we don't see these differences reliably in individuals—we usually have to group the data in order to see any patterns. And they're very sensitive to changes in experimental conditions and context. But in principle, it's possible that twenty, fifty, one hundred years from now, imaging devices may be available that could be used in these ways. So I ask the group whether this is a desirable goal for the future, and if so, how one might think about overseeing the development of such techniques.

Finally, I would just quickly mention that regarding the last of the seven sins—persistence, or the recollection of often-traumatic events that people cannot forget but would like to—there has been some progress. The research of people like Jim McGaugh, Larry Cahill, and others has shown that certain pharmacological agents can interfere with the development of these intrusive memories.

So now we are at a stage where people are starting to grapple with the use of such drugs. Is it a good thing, for example, to give a rape victim a beta-blocker, which has been shown in the laboratory to blunt the force of emotionally arousing memories? Or to administer such a drug to an emergency worker before he or she goes to a disaster scene that's likely to result in disturbing recollections? This is something that's much closer to reality than the distinction, through imaging, of true and false memories, and we'll need to deal with it sooner rather than later.

Question and Answer

MICHAEL WILLIAMS (Johns Hopkins University): I'm an intensive care neurologist, so I don't deal with Alzheimer's disease a whole lot, but certainly treatments for it are out there. How would you relate that to transience and your suggestion that we could have memory-enhancing drugs?

SCHACTER: I think that in cases of memory pathology—overt memory disorder—any ethical issues regarding an effective treatment are less pressing because there we have a disorder that's impairing the individual's ability to function and may eventually result in the loss of one's entire sense of self. So I don't see a big issue about using it, other than the usual kinds of questions about side effects and so forth. Stickier questions arise, though, if it's possible to improve people's memories from their baseline, or normal, level of function.

RONALD DE SOUSA (University of Toronto): Can you make clear your views on using the distinction between hippocampal and parahippocampal reactions, if indeed this results in a reliable test for true and false memories? I should have thought you'd want to use it in court to exonerate people much the same way we use DNA evidence.

SCHACTER: Let me first elaborate a little bit about the findings. We had conducted previous studies in which we had seen no difference anywhere in the medial temporal lobe between true and false memories in that word-list exercise. But in other behavioral work we found that you can improve the subject's recognition of words that really were on the list—and reduce the likelihood of the subject's responding to false lures—by making the presentation of the items a little bit more perceptually distinct. For example, you can present some of these words along with pictures, or tell the subject that it's important to link each word with the person who presented it. That way, when a word that really was on the list pops up, he or she may have a little perceptual tag in memory to enable the separation of true from false. And that is in fact what we found behaviorally and also neurally: the parahippocampal region seems to show stronger activation when perceptual qualities of the stimulus are maintained between study and test.

Now, coming back to your question about the implications of this. Given that the finding seems to hinge on a critical set of conditions being fulfilled, even if we saw distinctive results in every single subject, we couldn't control the real-world situation. We wouldn't know whether sufficient perceptual encoding accompanied the memories of interest. That would make me very wary about going into a courtroom with such a technique. It will be a long time before we're in any position to use it outside purely experimental studies.

Traumatic Brain Injury and Legal Responsibility

SUMMARY: Dr. Winslade described what he considered to be a major omission in the criminal-justice system: the failure to take the accused's frequent condition of traumatic brain injury into account. He cited two cases—one of fifteen death row inmates, and the other of an unfortunate young man rendered behaviorally incompetent by an automobile accident—both of which illustrated the need for law and medicine to work together to produce more appropriate, humane, and fair outcomes. He argued that while future scientific advances could potentially be of great value, the legal system has yet to begin implementing what has already become available.

WILLIAM J. WINSLADE: Following up on what Pat Churchland and Ken Schaffner said, I'm going to talk about being in control and out of control as a result of traumatic brain injury; I will also briefly discuss excusing conditions and mitigating circumstances as well.

It's well known that serious traumatic brain injury can cause all sorts of cognitive, personality, emotional, and behavioral changes. I included in the conference readings a fascinating and important article by

Dorothy Otnow Lewis, published in 1986, about a study done in 1984 with fifteen men on death row. They were the next fifteen people scheduled to be executed and were chosen for that reason alone.

Lewis did neurological, psychiatric, and psycho-educational studies of these death row inmates, and the only part that's relevant to my presentation today is the fact that all fifteen of them had suffered serious traumatic brain injuries. Nobody had taken their brain injury into consideration either in their defense or in their sentencing. The point of her article, which is quite convincing, is that this information *ought* to be taken into consideration in determining how to respond to someone who has committed murder.

I thought, "Well, surely this has been taken care of by now" when I was invited two years ago to give a talk to the National Public Defenders Meeting. Among the 800 public defenders from around the country who were present, I found only a handful who had ever heard of this article, much less taken into account the effects of traumatic brain injury on the people they were defending on murder charges. The law lagged way behind even clinical scientific investigation, not to mention the more sophisticated neuroscientific work of the laboratory.

The law has to become much more aware of recent scientific developments. After all, DNA testing (which has been mentioned at this conference) has caused a revolution in the law because now we can find out whether somebody did or didn't do what they were accused of doing, based on reliable scientific evidence. Admittedly, the science of the brain is a much more complicated matter. Still, some legal

Dr. William Winslade, University of Texas Medical Branch.

awareness of what scientists do know could make a big difference in the justice system and for affected individuals and families. Here is an actual case:

John was a normal guy who in the early 1970s was in his early twenties. He had a job and was working his way through the different phases of his father's company. Actually, he was about to inherit the company. But then he had a horrible car accident. His girlfriend was riding with him, and she was killed. John himself was in the hospital for eight months before he recovered from all his physical injuries, and when he was released from the hospital, he had changed. John had been a normal guy and now he was a paranoid.

In fact, he became a fulminating paranoid. John, who had moved back into his parents' home because of all of his physical handicaps, believed that there was a Czechoslovakian conspiracy to kill his father. One day he was at the drugstore with his mother when she was picking up some medication for his father. The pharmacist said to her jokingly, "What are you going to do with all that rat poison?" John decided that his mother was part of this conspiracy; he went home and promptly shot and killed her. Until that horrible incident there had been no conflict in the family; John had gotten along well with his parents.

John was evaluated by a number of prominent psychiatrists in Southern California at the time. He didn't have a trial; he was just sent directly to the mental hospital. He was diagnosed as an insane schizophrenic. His father persisted in trying to get him released from the psychiatric hospital because he didn't believe that John was truly insane. Obviously something was wrong, but it wasn't that he was mentally ill. After about five years, his father did get him released, but then John got in trouble again. In response to an argument with a neighbor, he fired a rifle in the air. He was sent back to the mental hospital, where he remained for twenty-eight years.

Over time, many of John's psychotic behaviors resolved, even without effective psychiatric treatment. When I and a group of other people evaluated him, it was clear that nobody had ever before evaluated his brain injury—not at the beginning and not during the twenty-eight years he had been in the hospital. And now that he was about to be released, the State of California first wanted him to pay an $800,000 bill for his long, though involuntary, hospitalization, during which he did not receive appropriate medical care.

The "happy ending" is that we negotiated to eliminate most of the charges, but John might have benefited far more—and avoided much of the long and perhaps largely unnecessary confinement in that hospital—if there had been a careful neuroscientific evaluation. It might have been possible to determine whether his brain was out of control because of the brain injury or because of mental illness, or whether he was just mad at his

If brain science...can give us better diagnostic evaluation, better therapeutic efficacy, or better predictive power, these outcomes will significantly change the way we look at criminal justice.

mother and acting out a Freudian fantasy of homicidal delusion. Who knows? But the point of these examples, the death row inmates and John, is to say that if brain science, including neuroimaging and other technologies, can give us better diagnostic evaluation, better therapeutic efficacy, or better predictive power, these outcomes will significantly change the way we look at criminal justice.

For example, if we had properly diagnosed a brain-injured man on death row, there might have been the mitigation of punishment that Ken Schaffner talked about and that H.L.A. Hart recommends for somebody who may well have been out of control. And a thorough brain analysis might also show that a person who claimed to have a brain injury didn't have one at all. So it isn't that better assessment is just going to free people from responsibility; it will help us learn whether they are capable or incapable of controlling their conduct.

The law has a cognitive bias, and its most extreme form is the M'Naghten rule. According to this doctrine you can't be found "not guilty by reason of insanity" unless you don't know that what you're doing is wrong. Andrea Yates was found guilty under the M'Naghten rule because it was deemed that she indeed knew right from wrong and knew that her conduct was wrong. Clearly this was someone in emotional chaos and out of control. It may be a much more difficult task to show that somebody should be exonerated from legal responsibility because of brain injury or other brain dysfunctions. But at least there would be a basis for trying to assess, on the spectrum of control, how much control a person actually had over his or her conduct.

As we move into the more troubling area of predicting behavior, even there we can benefit a great deal from successful work in the neurosciences. Take John, for example. It's quite possible that we would have found, in spite of his condition, that this was someone whose behavior required that he be incarcerated. But instead of sending him to a psychiatric hospital where he didn't get the treatment or the rehabilitative support that he needed, we might very well have been able to address his underlying condition and modify the management and the incarceration.

When I interviewed him after his twenty-eight years in the hospital, it was clear that he wasn't somebody who was ready to just go right back out on the street; his ability to control his conduct had been permanently impaired as a result of his brain injury. So he had to have a sheltered living environment, and he had to have follow-up care. The mental health professionals of the state psychiatric hospital realized early on that John did not belong there. The irony was that he actually had the resources to get the services he needed if he could ever have gotten out of the institution, but the state was trying to keep him from leaving simply because he hadn't paid the bill that he shouldn't have had to pay in the first place.

Analogous to DNA testing, to the extent that neuroscience can provide reliable and objective evidence, this testing method is going to make the criminal justice system both fairer and more appropriate in its response to people who commit crimes. I can't do anything about the adversary system, which distorts anything that is going to be introduced into a criminal trial, especially one that's highly emotional. So if you think about John killing his mother, or Andrea Yates killing her five children, it's very likely that all sorts of extraneous factors will enter into the decisions of a jury or judge as to whether or not someone is legally guilty.

But at least one of the things we should strive for in thinking about the implications of neuroscience is to improve the criminal justice system—make it more appropriate, fair, truthful, and honest—notwithstanding the impediments that the adversary system will inevitably impose on it. Of course, there's always a danger of premature intervention or premature scientific claim. But in my experience, at least in this area of traumatic brain injury, the law has been very slow to profit even from knowledge that's already out there. So I'm hoping that genuine scientific advances will help it toward better insights in the future.

At least one of the things we should strive for in thinking about the implications of neuroscience is to improve the criminal justice system—make it more appropriate, fair, truthful, and honest.

Question and Answer

STEVEN HYMAN (Harvard University): Let me just ask for a clarification. One of the things you said about the John case, which was very good, was that it sounded like this unfortunate soul wasn't well evaluated and that you would have treated him very, very differently—more humanely and appropriately—and when he got out things might have gone better. But I think we have to worry about our ability to actually evaluate these patients. I would argue that we need a good dose of humility—that we're really quite far from being able to map things like intentionality onto almost any MRI or set of neuropsychological tests.

WINSLADE: You want me to be more humble?

HYMAN: No, no, no. I'd just like to get your sense of where we really are in terms of any kind of objective test that could exonerate somebody.

WINSLADE: I completely agree with everything you said. I was pushing, speculatively, toward the future. In John's case and in the case of Dorothy Lewis's studies, the point was that nobody was even looking at traumatic brain injury at that time. And what troubles me is that nobody's looking at it very much now either, even though there are things that can be done from a clinical evaluation point of view that would be relevant, though not decisive. I think it would be very interesting to explore the next step—to combine clinical evaluation with whatever available evidence there is from more sophisticated forms of neuro-imaging.

If we found that a huge percentage of our incarcerated individuals truly were brain injured, what would we do with that information?

MELANIE LEITNER (AAAS Fellow): Here's one question that's hard and another that's even harder. How do you look at drug addiction? Is it an illness that should be considered when evaluating individuals in the legal system? And the second question is,

What if we do become more effective at detecting brain injury? It's a double-edged sword, because while it might make us more likely to excuse an individual, it means we're also less likely to rehabilitate that person, even assuming that our criminal system is rehabilitative. So if we found that a huge percentage of our incarcerated individuals truly were brain injured, what would we do with that information?

WINSLADE: With respect to your first question about drug addiction, I don't have enough expertise to say anything about it. With respect to brain injury among people in prisons, they do in fact have a very high prevalence of brain injury, and I'm not sure what we should do about that. But we haven't even *begun* to address it. The condition is certainly relevant to the question of whether or not individuals can control their conduct; and from the criminal-justice system's point of view, ability to control your conduct is a necessary condition for being found legally responsible. Still, I think we have a lot to learn, following Ken Schaffner's point about the difference between excusing and mitigating circumstances and H.L.A. Hart's wise observations on how to implement that knowledge. As a practical matter, it's a very difficult task.

Neuroethics and ELSI: Some Comparisons and Considerations

SUMMARY: Professor Greely compared some of the likely ethical, legal, and social implications of neuroscience with similar effects, already being studied, of genetics. He discussed three subjects in particular: prediction, human cloning, and determinism/essentialism. Neuroethics issues will arise in each area, some of them similar to those of genetics, and others unique to the brain. Outcomes will depend heavily on future scientific results, as well as decision makers' belief systems. If advances in neuroscience are as far-reaching as currently expected, neuroethics will be an important field.

Dr. Henry Greely, Stanford University.

HENRY T. GREELY: As we have heard, "neuroethics" is a new term. A wordsmith can launch a new word on the ocean of the language, but whether it sinks immediately or flourishes is largely beyond his control. I want to discuss the question, What is the likely future of neuroethics? Will it become a field or subfield of bioethics or will it be just one more area in which existing disciplines explore the social implications of new technologies? A good precedent for comparison is the field that explores the ethical, legal, and social implications of human genetics (ELSI), an area that has been pursued through well-funded investigations for the past decade.

What lessons can we draw from ELSI for neuroethics—the study of the ethical, legal, and social implications of neuroscience? Most of today's discussion has focused on issues of free will, determinism, and criminal responsibility. Those are tremendously interesting and important questions, but neuroscience may affect society in many other crucial ways that are not captured in those big questions. In genetics, one can divide such issues into six broad categories: genetics and identity, the effects of genetics in revealing the past, the effects of genetics in predicting the future, the manipulation of genes, the ownership and control of genes and genetic information, the cultural effects of genetics, and the consequences of genetics for our culture.

With the possible exception of the first—identity—you could substitute *neuroscience* for *genetics* in all those categories and have meaningful sets of questions. I want to demonstrate this with three examples—prediction, human cloning, and issues of determinism or essentialism—before ending with some comments on whether anything like an ELSI program in neuroscience is likely to happen.

The first issue, prediction, is actually very easy to talk about in neuroscience. Neuroscience and genetics not only aren't entirely separate in this report, they're largely overlapping. One of the best examples of predictive genes is the alleles (genetic variants) that cause

Huntington's disease. Neuroscience and genetics intersect in this neurological disorder. In thinking about how people might use predictions made from neuroscience, it's useful to distinguish between the prenatal and postnatal predictors.

Prenatal prediction would use information obtained from neuroscience to aid in decisions about whether to abort a fetus, to implant an embryo (through preimplantation genetic diagnosis), or to mate with a particular person. Such choices would be aimed at producing progeny that have normal capabilities, "supernormal" capabilities (in cases of enhancement), or "subnormal" capabilities (as in the recent case of a deaf couple who decided to have a deaf child). These issues of prenatal choice are affected by whose choice you're talking about. Is it the choice of the parents? Is it the decision of the state? Or, in an area that I think is underanalyzed, is it the choice of parents who are heavily influenced, one way or another, by the state? To the extent that neuroscience can tell people something meaningful about their likely progeny—possibly at the embryo stage but more likely at the fetal stage—the kinds of issues that we've confronted in genetics will have to be confronted in neuroscience as well.

To the extent that neuroscience can tell people something meaningful about their likely progeny... the kinds of issues that we've confronted in genetics will have to be confronted in neuroscience as well.

Similarly, questions of postnatal prediction will be important. We think of this most commonly in human genetics in the context of highly penetrant late-onset diseases, such as, for example, Huntington's disease. As far as we know, if you carry one copy of the genetic variation that causes Huntington's disease, the only way not to die from that disease is to die first from something else. It is an extremely powerful allele. In that situation, there may be some advantage to self-knowledge in improving your ability to plan your future. Also, one hopes for eventual treatment or preventive interventions, though none exist now. But this knowledge comes with some costs. People worry about the implications of genetic tests for employability, insurability, relationships with family members, and the psychological well-being of the person who decides to be tested.

The same problems can arise from predictions obtained from neuroscience. To use Huntington's disease again as an example, a neurological examination that makes this diagnosis has effects similar to a positive genetic test. That diagnosis, like the genetic test, provides information that has consequences for future employment, insurance, family relations, and personal happiness. In addition, neuroscience predictions may affect areas not as heavily discussed in the genetics field including prominently the consequences for how people are to be educated. Predictions could also lead to the increased surveillance of people considered dangerous or even to the preventive detention of those believed, on neuroscience grounds, to pose threats to themselves or others. The predictive side of neuroscience is likely to have major effects on society, and people need to think about what, if anything, we should do to forestall or encourage them.

I raise the second issue—cloning—mainly because of its timeliness. Sometime in the next two weeks the U.S. Senate is supposed to decide whether it will become a federal felony, punishable by ten years in prison and a million-dollar fine, to clone a human embryo. The cloning issue will of course have consequences for neuroscientists. Researchers will want to use cloned cells for a variety of research applications; clinicians working to treat neurological diseases will be interested in the possible therapeutic uses of cloned cells.

But neuroscience is intimately bound up with the cloning and stem cell debates in other ways. At legislative hearings on cloning, the most effective testimony comes from disease organizations focused on neurological issues. People with Parkinson's disease or with spinal cord injuries are compelling witnesses for the importance of further research. The reality of their hopes—and to some extent, their political power—hinges on what neuroscientists say about the promise of these techniques. In another respect, neuroscience might say something to help at least a few people decide what they think about the moral status of the embryo through revealing the development of the neurological system.

The third area is determinism or essentialism. These two terms are sometimes confused, although they are importantly distinct. Happily, I can cite a geneticist as an example of each of them. James Watson was famously quoted as saying, "We used to think that our fates were in our stars. Now we know that they're in our genes." That's a foolish thing to say, but it makes a nice example of genetic determinism—the belief that everything about your future is contained in your genome. As Ken Schaffner already noted at this conference, it has become clear through genetic research that, for most people, genes in fact are not all that determinative.

For some people, however, they are. If you are born with two copies of an allele for Tay-Sachs disease, your future is heavily determined—you will die, and die young. If you are born with one allele for Huntington's disease, your future demise is also determined, although forty, fifty, or sixty years later. Most of us, however, are not born with alleles that have such a strong determinative role in our lives.

If you are born with one allele for Huntington's disease, your future is also determined, although forty, fifty, or sixty years later.

Whether neuroscience predictions turn out to be that determinative or not is still unknown; we don't yet know what the neuroscientists will come up with. My guess, based on the precedent of genetics, is that in a few cases neuroscience will lead to incredibly powerful predictions. In others, there will be moderately powerful predictions; and in most, there will be either weak predictions or none at all. But whether and to what extent determinism turns out to be an issue in neuroscience depends heavily on how the science develops.

Essentialism is a more interesting issue in neuroscience than in genetics. Walter Gilbert once said, "In a few years we'll be able to put your entire genome on a compact disc." And holding up a compact disc to the audience, he added, "See, this will be you." That my genes will be me—that I am my genome—is ridiculous, given all the influences of environment, chance, and time on who I am and what I'm thinking about.

I am more than my genes. The genes are an

important part of me, but I can be certain that they are not my essence; they are not my soul. When we shift that notion to the neuroscience area, though, I am not so confident. Is my consciousness—is my brain—me? I am tempted to think it is. But of course, all of us in this room, coming from backgrounds that emphasize intellectual effort, may be biased on this point. Still, it is more tempting to think that our thoughts, our consciousnesses, our brains are more ourselves than our genomes ever could be. That has several consequences.

Some years ago Stanford scientist Irv Weissman, a stem cell specialist, created something called the SCID-hu mouse, a mouse with severe combined immunodeficiency that had, in place of its own non-existent immune system, a transplanted human immune system. In that way, one could study the human immune system in vivo. Dr. Weissman is now interested in doing the same thing with human neurons, to create a human-neuron mouse—one whose own neurons would have died off either prenatally or shortly after birth and would be replaced with human neurons. In that way, he could study human neurons, in vivo, in a laboratory animal.

I have been one member of a five-person group studying the ethical implications of this tremendously interesting effort. As part of our work we have clearly realized that talking about a mouse with a human immune system—with human bone marrow—feels quite different from talking about one with human neurons. Because most of us would likely agree that the brain is somehow more involved in our humanness than is our bone marrow. Is this reaction neuro-essentialism?

Another side of neuro-essentialism, one that might have more bite in the real world, is defining death. One's view of neuro-essentialism could have some distinct consequences, for example, in deciding how to treat anencephalic children, those born without a cerebrum. Whether you think they are living or dead humans or living but nonhuman, could make a great difference in how they should be treated.

Another side of neuro-essentialism, one that might have more bite in the real world, is defining death.

These have been just a few examples of some of the issues in neuroethics. There are plenty more where they came from. And many of them have nothing to do with perennially interesting questions like free will or criminal liability. Will they be studied by a field called "neuroethics?" People like Bill Winslade, without using the term, have been doing neuroethics for twenty years and more. They will continue to do this work, but will it be recognized as a distinct field? Will it become a major area of research? That's impossible to know, although on this question the comparison with the ELSI program is daunting.

There is no equivalent in neuroscience to the Human Genome Project. One could argue that the driving force behind the ELSI program and its commitment of tens of millions of dollars to study ethical, legal, and social issues was a cold-blooded attempt to buy off political opposition to the Human Genome Project. By saying that the project would itself spend a lot of money—3 to 5 percent of its total budget—to study these issues, it quieted some of the project's political opposition. Neuroscience will not have an equivalent to the Human Genome Project; it does not provide a similar, conceptually simple target. Therefore, this motivation for funding neuroethics will be missing.

On the other hand, some of the social and ethical questions raised by neuroscience are likely to be even more interesting and important than those raised by human genetics. To the extent that neuroscience raises such questions for society, somehow research into those questions—and research funding—will follow. Basically, the future of neuroethics is up to the neuroscientists. If they build it—if they find things that have social implications—neuroethics will come.

Question and Answer

STEPHANIE BIRD (MIT): Because predicting something about your liver is quite different from predicting something about your brain and how you're going to see the world, I'm wondering about how we might, as a community, be especially thoughtful and effective in presenting our neuroscience findings to individuals. This is especially critical in that some brain-related predictions could well become self-fulfilling prophecies.

GREELY: That has been a tremendously important issue with respect to genetic testing. The genetic conditions that are very highly penetrant—where people with a particular genotype get the disease close to 100 percent of the time—turn out to be pretty rare. For example, though there are at least three highly penetrant genes that, when mutated, lead to Alzheimer's disease (typically, early onset), they account for maybe 1 percent of people with Alzheimer's.

More common are generic associations like that of the ApoE-4 gene, which doubles or triples one's risk of Alzheimer's disease but doesn't produce even a 50 percent total risk of developing the disease. Presenting such information, and knowing when it's sensible to present it, poses a difficult problem for counselors, physicians, and others who talk to people about possibly getting that test. These professionals have to be quite sophisticated and empathetic in providing information, and allow the patient to make an informed decision about whether taking this test is going to make sense in his or her circumstances.

We have thought about that more in the genetics area than in other parts of medicine. Nobody worries about giving their informed consent before they take a cholesterol test, although that test might very well be influential in predicting one's future. And I suspect the same is true of many neurological conditions and many psychiatric and psychological tests.

DAVID PERRY (Santa Clara University): I am particularly interested in the connection you drew between brain functioning and moral status. So I wonder why, in your state commission's report on cloning, you folks recommended that therapeutic cloning be permitted, but only up to the formation at fourteen days [when the pre-embryo stage after fertilization is thought to end] of the "primitive streak" [a band of cells from which the embryo develops]. That's the only part of that report I was disappointed in, because you went no further than all of these other commissions. Yet it seems to me that at fourteen days you're still very far from an entity that is capable of consciousness, and by stopping there we'll likely be unable to investigate why embryos don't implant, for example, or how brain abnormalities like anencephaly develop. Can you comment on that?

Before cells begin to differentiate into their functions, it seems very hard for anyone to argue that there is the remotest chance that sentience exists in that small ball of cells.

GREELY: The commission to which David refers is the California Advisory Committee on Human Cloning, on which I serve. It reported to the legislature in January of this year, recommending a continuation of California's first-in-the-nation ban on human reproductive cloning but also approval for nonreproductive cloning, subject to regulation. One of the regulatory constraints we recommended was that research should not be permitted with embryos past the appearance of the primitive streak. That qualification was driven, I think it's fair to say, largely by two things: a very strong desire to have a unanimous report, and the fact that it was a fairly straightforward albeit very conservative place to stop, at least for now, based on current knowledge.

Before cells begin to differentiate in their functions, it seems very hard for anyone to argue that there is the remotest chance that sentience exists in that small ball of cells. Even past the development of the primitive streak and the first real commitment of cells to different functions, my own guess is that any neurological functioning will not come for many days and weeks. But fourteen days was a good, easy, clear stopping point for now, based on our current understanding. We did not mean that fourteen days would always be the limit; that limit could be changed in the future based on new understandings that would likely come from neuroscience.

JUDY ILLES (Stanford University): Given our genetic and neuroimaging testing capabilities now—the technology to predict for deafness or preselect for dwarfism, and possibly to survey for individuals who may be aggressive or have a predisposition for drug addiction—are we obliged to revisit or redefine what is normal?

GREELY: Perhaps. *Normal* is a very slippery term. I have always liked to think of normal as a distribution and not a point. But to the extent we begin to understand the reasons *why* people are outliers in one direction or another, we may want to rethink the meaning of normal. Frankly, I don't know the answer, though I think that's a great question for the first generation of neuroethicists. Did I duck that question adequately?

Luncheon Speech

No-Brainer: Can We Cope with the Ethical Ramifications of New Knowledge of the Human Brain?

SUMMARY: Dr. Caplan's answer to the title question was yes, provided we begin grappling now—start setting standards and forming basic policies—related to ethical issues raised by advances in neuroscience. He went on to develop one such issue: Should we try to use knowledge of the brain to improve ourselves? If a drug or an implant, for example, could enhance our memory or teach us French, should we use it? He maintained that we certainly should, reflecting the time-honored human desire to make ourselves, and especially our children, better. Dr. Caplan acknowledged societal inequities in advantage and access but argued that our goal should be to reduce unfairness, not eliminate beneficial options. He cited other common objections and rebutted them in turn.

William Safire.

WILLIAM SAFIRE: Our box-luncheon speaker, Arthur Caplan, is a man who cannot be boxed in. He's the director of the Center for Bioethics at the University of Pennsylvania and the Hart Professor of Bioethics at Penn's School of Medicine. He also holds professorships in philosophy and psychiatry. He turns out a lot of copy. He has contributed more than 400 articles and reviews to professional journals, which is more than I do in a year. He has written several provocative books, with some good titles like *Am I My Brother's Keeper?* And—my favorite—*If I Were a Rich Man, Could I Buy a Pancreas?*

Before he joined Penn, Dr. Caplan served as director of the Center for Biomedical Ethics and professor of philosophy and surgery at the University of Minnesota. He was a distinguished lacrosse player. He is also known as the ethical consultant to the Pfizer Corporation on its development of Viagra. He points out to me, rather acerbically, that he was the consultant and not anything else. He brings us today what he calls a no-brainer (a term coined in 1970). His talk is "Can We Cope with the Ethical Ramifications of New Knowledge of the Human Brain?" If the answer to that question is no, he is going to look at a very depressed audience.

ARTHUR CAPLAN: I believe we *can* cope with the

ethical ramifications of new knowledge of the human brain. I believe this so strongly that I will try to defend the position that we should use the new knowledge the brain sciences are providing to try to improve, enhance, and otherwise move toward optimization of our brains.

In other words, I want to take the provocative position that even before we know how to do it, we should anticipate that we will want to improve our brains. A lot of information is coming from the brain sciences—in such areas as neuroscience, radiology, psychiatry, and behavioral genetics—that will give us opportunities to think about how we might modify and design ourselves.

When issues of where to go with new scientific knowledge are raised, we discuss these questions in a somewhat academic mode that's separate from where the culture might be on matters of science. Where do you think the American people actually get their most direct exposure, for instance, to the world of genetic testing? What avenue provides them with that knowledge? Is it a familiarity with Huntington's disease testing, with informed consent and genetic counselors? Is it BRCA1 testing? I don't have a study to prove my point, but I'll tell you what I'm convinced is the source of most Americans' views about genetic testing—it is television programs such as the *Jerry Springer Show, Montel,* and *Maury Povich.*

Dr. Arthur Caplan, University of Pennsylvania.

These shows love to do on-air

paternity testing. The crass world of daytime TV says the way genetic testing works is that you get a genetic test, the results are presented on national television, and you have to make decisions about an intimate area of your life in a few seconds as a crowd hoots at you.

That's a model of genetic testing that is very well understood by most Americans. Advances in neuroscience could go the same way. I'll make a prediction that in ten years there will be a show called *My Brain Made Me Do It.* Someone will lie in a CT scanner that quickly takes a picture of his head while the host asks, "Guilty or not guilty?" Then, head image in hand, a test expert will come out and say, "You know, his amygdala is kind of big. I think he did it." The audience will then hoot.

And that's the way our culture now deals with advances in the frontier of biomedicine. There may be times when we bemoan it and wish to isolate ourselves from it. Those of you doing academic work may not want to intersect at all with that grubby arena of entertainment, product marketing, and all the rest. But there it is, and out of that world will come much of the public's knowledge of the brain—sometimes with qualifications, sometimes not. So as I'm about to mount a defense of the idea that some information in the brain sciences should be used to try to improve and enhance human beings, I do so despite having a good idea about how things can be misunderstood, misapplied, and exploited in our culture. Nonetheless, it seems to me that if we look out and start to see what the revolution in brain sciences is beginning to accomplish, we have a set of immediate issues before us that those in the field of genetics did not adequately address. We should put some of these ideas on the table and talk about them.

For one thing, there was never sufficient discussion of standards for guiding genetic testing: who can offer it, how accurate it should be, and whether counseling is needed. Isn't it important to begin talking about the formulation of standards—on accuracy, for example, or competency—in basic brain-imagery work?

Another key step in managing new technology is that if you're going to have people stepping into court soon—and I'll bet someone out there is writing down the remarks that Bill Winslade made to us earlier, about looking for traumatic brain injury and getting that defense going for their client—it would also be useful to say what standards are in place for psychiatrists' or neurologists' assessment.

Are there any standards yet for privacy and confidentiality of neuroscience clinical information? Actually, no.

Are there any standards yet for privacy and confidentiality of neuroscience clinical information? Actually, no, there aren't, and that is not a good situation to be in if we're going to see this revolution move forward. In one case, doctors have come to me and reported that, in trying to assess different types of brain-scanning technology for patterns and characteristics that might identify someone as having a mental illness, they also have in front of them their subjects' history, patient chart, whether they've been arrested, whether they've served time in jail, and so on. These researchers report all kinds of correlations between lifestyle patterns and behavior—sometimes associated with ethnicity and race—and those they observe of their subjects' brains. Now this may all be fallacious reasoning, it may be loose association, it may present *post hoc, ergo propter hoc* kinds of problems, but the fact is that this information is flying all over the place. Coping with the neuroscience revolution means taking privacy seriously and doing something about it.

I might also mention something that seems to have been forgotten in the current battles over stem cell research and cloning of stem cells. About ten years ago, some veterans of earlier bioethics wars will remember, we had the fetal-tissue wars. In the United States, the whole debate was around the fetus and its link to abortion. But in Sweden, when they established their codes for fetal-tissue research, they did not focus on abortion. Instead they worried about altering personal identity through a brain tissue transplant. Swedish policy insisted that no scientist implant in anybody's brain anything more than a clump of cells. All brain tissue has to be disaggregated. The Swedes were terrified of personal-identity modification by transplanting fetal tissue into a brain. Did they know what they were talking about? I have no idea, but perhaps they envisioned that a critical mass of fetal tissue in the dopamine-secreting region of your brain would turn you from a friendly Swede into a xenophobic American.

But whatever miracle of transformation they had in mind, at least they realized that when you muck around in someone's head—whether by deep implants to treat epilepsy, or delicate microsurgery to try to modify a particular mental problem—you may be threatening someone's sense of who they are. You may be modifying them in such a way that their personal identity is changed. These are very real and important questions, and I do think we need to be grappling with them. Right now, in the short run, some very simple, clear, obvious policy issues must be targeted for discussion, resolution, and recommendation.

But let's presume we can make the world safer for neuroscience. The one issue I want to spend the rest of my time addressing is this: Should we try to use knowledge of the brain to improve ourselves? This was hinted at in a previous presentation when Professor Schacter essentially said, "I'm a little uncomfortable about moving past baseline here. I've looked at a lot of types of memories and I understand something about these mechanisms and it's one thing to repair and treat but perhaps an entirely different thing to enhance and optimize." I think it *is* a different thing to enhance and optimize, but let me throw caution to the wind (a very unusual stance for me) and ask, why *not* go for enhancement?

Should we try to use knowledge of the brain to improve ourselves?

Let's say I've got an interest in learning French, and that a doctor has a "mind machine," or a pill, or a tiny implant, or a bit of a brain of a Frenchman. One way or another, he's able to get this into me and I don't have to spend the entire summer going to Berlitz or some other tutorial course. I'm going to learn French in minutes because I've got a kind of French "mind meld" opportunity through some manipulation of my brain (for those of you who remember Spock and the first *Star Trek*).

Why shouldn't I do this? What's bad about this?

Why wouldn't I want to enhance, improve, invigorate, optimize my pathetic mind and try to wind up with something better than what I've got? I don't claim to have a disability. I don't claim to be beyond the average in a normal distribution. I just don't like taking a long time to learn French; I want to learn as fast as I can.

Now, I'm thinking about this a lot because I still have a 18-year-old in my house who is thinking about going to college. And as I watch his—how can I generously describe this?—evolving mind grasp the challenges before him, I note that many of his fellow students are signing up for something called the Kaplan course to improve their SAT scores. And I see very little ethical fretting on the part of their parents, other than whether the course will work. If I pay this money, will Kaplan guarantee me a twenty-point boost in the test?

I'm not here to argue about whether SAT tests are good or whether the merit system that we supposedly have in place makes any sense. All I know is that here's this test, and my son's entire school is berserk trying to find ways to maximize their performance on the test. And no one asks, What the heck is going on here? Have these people thought about the fact that they're trying to move their kids—normal kids, some of them—outside the baseline? That success in the Kaplan course could shift them from where they are on the curve? If they have thought about it in that way, they're likely all for it. The employ of Stanley Kaplan and his ilk as stimuli to engineer and improve their kids is seen by kids and parents alike as, if you'll pardon the expression, a no-brainer.

I also observe that my son goes to the Germantown

Friends School. I pay about $15,000 a year to send him there. Do you know what he has? A huge advantage. He is a privileged kid. If I go to a poor neighborhood and say that I send my kid to this school, they don't say to me, "You should be ashamed. You're giving him an advantage." What they say to me is, "I wish I could give my kid that advantage." Now, it may not be a fair system—personally, I don't think it's a fair system at all—but the notion that we should strive to improve, enhance, optimize, and make our kids better off than we were is pretty deeply engrained in every moral system that I know of, religious or secular. I doubt there are any ethical systems that say, "Take your kid and make him or her worse off. That's your duty; do what you can." True, parents may manage to do that, but that's a different thing from thinking it's the right thing to do.

Without a doubt, they're not trying to repair anything when parents want to boost their kids' test scores. They are just trying to improve them. The cultural message is that this is not a bad thing to do. It is good. Why then do we recoil at the idea that if we had knowledge about the brain, even imperfect knowledge, we should use it to try to make ourselves or our kids better?

Let me give you some examples, beginning with the armed services. If you have an ability to scan people's visual field inside their brains and they're going to fly a very expensive stealth bomber, you're interested in knowing about their reflexes and peripheral vision. Brain scanning is believed to be able to tell you a little bit about that. It may give you, let's say, a 20 percent chance of being able to detect someone with exceptional peripheral vision. If you go to the Defense Department and ask if that's something they're interested in, even at that poor level of testing, they'll likely say, "Given the fact that the plane cost $2 billion, we're real interested. Begin testing. We want the best pilots possible, and if we're in error about this, okay, but if we're a little better at picking better pilots, great."

Consider the drug you may have been reading about that allows people to stay up longer. I had breakfast this morning with the head of a big financial firm, and he wants it—tomorrow! I said to him, "It could have side effects," and he answered, "Yeah, good." He wants to stay up longer and turn more income, and perhaps some of his employees will follow suit. We all know that in other areas, like the so-called lifestyle drugs—whether it's Viagra or things to repair our wizened faces, small mammary glands, or furrowed brows—we have people who are trying to improve appearance or trying to improve performance.

We have the notion that you should earn what you get, and that if youtake a pill or use a surgical scalpel or drop in an implant, somehow you've cheated.

I think we're a little puritanical about this in the United States. We have the notion that you should earn what you get, and that if you take a pill or use a surgical scalpel or drop in an implant, somehow you've cheated. But in a deeper sense, do we really cheat if the outcome is that we've made somebody perform better, able to achieve more, have greater capacities? Here are some of the arguments I've heard that say we do cheat.

For one thing, improvement would supposedly be unfair. Some people would get it and some people wouldn't. Well, that's certainly true, but the reason I told you the story about Stanley Kaplan and the private high school is that it's unfair now. That doesn't make it right. But the solution is to make it fair, not to do away with the improvement. We should try to insure that *everybody* can go to a good school. I wish everybody had access to whatever tricks we can find to increase memory or allow people to get by with a little bit less sleep. That they don't have equal access is a problem, but it's not one that is inherent in improvement. Nor is unequal access an argument against improvement; it's an argument against inequity.

A second objection is that the quality of human beings might be threatened if you start to let some people become advantaged. The quality of human beings does not presuppose, however, biological equality. It is instead a claim about moral worth that goes beyond particular attributes, properties, and behaviors. It is, if you will, a normative stance about how you want to treat human beings. But it's simply fallacious to say that our notion of equality depends on having some kind of regression to the norm of biological existence.

A third argument is that it's wrong to improve because it would be unforgiving: if we do this and we improve, then the disabled—the different among us—will be disadvantaged.

Well, there is always a risk of discrimination, but the same problems of disability and difference confront us now, and there's no reason that people who are different should be foreclosed from using options about brain intervention to improve or enhance themselves. The access should be there to make it fair. And there's no reason to think that trying to make ourselves have a better memory is going to make us feel worse about people with Alzheimer's. We can do what's wrong in terms of dealing with people with Alzheimer's, but that's not an argument that it is not right to try to improve one's memory.

This argument about what we can or cannot do to design ourselves is made by people who wear eyeglasses, use insulin, have artificial hips or heart valves...

Fourth and last objection: It's unnatural. This is one that my friend Leon Kass has been promoting a great deal—that it's wrong to muck with our nature. He and some of his acolytes, such as William Kristol and Charles Krauthammer, as well as Francis Fukuyama, suggest that if we start to muck around at improving and enhancing ourselves, we're going to become "posthuman." This argument about what we can or cannot do to design ourselves is made by people who wear eyeglasses, use insulin, have artificial hips or heart valves, profit from tissue or organ transplants, ride on airplanes, talk on phones, and sit under electric lights. What are they talking about? Are we posthuman if we ride but don't walk? We might be less healthy but posthuman? I don't see an argument here that says there's a natural boundary or limit that tells us that our nature is defiled by technology.

So in sum, I don't think the arguments are persuasive about why we shouldn't try to improve ourselves. I, for one, await the day when my neuroscience friends give me some ways to do it.

Question and Answer

WILLIAM SAFIRE: Let me pose the first question, regarding something you didn't address with the four straw men you set up and knocked down. Here's a fifth. With advantage comes control. Let's say that the fairness issue arises and the government says, Why should some rich kids get intelligence enhanced by twenty-five points when poor kids can't afford it? So the government decides to supply a prescription drug or implantable chip for intelligence enhancement, as long as recipients accept a modicum of control: the drug or chip will also produce a tendency toward liberal voting. Now in China, they could easily control a billion and a half people in this way. Aren't you leading us, then, into an Orwellian world?

CAPLAN: Some Americans are obsessed with the notion of vast government power that, among other things, could extract political control as the price of improvement. The America that I live in only goes as far as extracting kickbacks as the price of filling potholes. I'm more worried about a slightly different kind of situation: I might be a person who has a susceptibility to creating a child who is deaf or has some other form of disability. And my neighbors come to me and say, "You know, the responsible thing to do is take the drug, take the implant. Otherwise, you're not a good person."

No one passed a law in the United States, and there's no government edict, that says women over 35 must have amniocentesis. Yet every place I go, whatever group I talk to, people say they think it's irresponsible to not do that. The culture of improvement in a capitalist society can be very compelling. I do worry about what government could extract as a condition of entrée, and I concede that it's an issue to manage. But I worry more about what our cultural pressures could do because—remember—I still want us to choose, and I'm not saying we each have to choose in any particular way. So I'm nervous about how we're going to control that heavy pressure to say, "You're unethical if you don't improve yourself or if that kid isn't made better."

STEVEN HYMAN (Harvard University): Aren't we in a feed-forward vicious cycle in which it's continually the "haves" who get to enhance their kids? If you happen to be well off, you send your kids to the best high school, they get the Kaplan course, they go to privileged universities, they do better. Now we're going to enhance them pharmacologically, and they're going to be stronger, more attractive, smarter. I mean, one could really spin a dystopia here, despite your cavalier dismissal of what Bill called an Orwellian world.

SAFIRE: The provost of Harvard is knocking elitism?

HYMAN: Just being devil's advocate.

CAPLAN: We definitely need to think long and hard about the class-based divisions we could open further by, if you will, accelerating biological or neuroscientific differences. There are real equity issues here, and I don't mean to put them aside. I'm just saying the technology gives us abilities and capacities that would be useful, and we have to figure out how to make sure that access to it is fair. But if we're going to say at the same time, "No movement until everybody's the same!" we've got another big problem, because we're living in a class-based dystopia right now. You know what the admissions look like at Harvard, and I know who's taking tennis lessons and who's going to summer camp while we also know who's living in West Philly and kicking a can down the street for entertainment.

It's not equitable now, so it would make no sense to argue that everybody has to have exactly the same opportunity before new benefits could be offered. But given the power of the technology we're talking about, we'd better close that gap somewhat. Similarly, we'd better think long and hard about the gaps that develop between us and the Third World.

It's not equitable now, so it would make no sense to argue that everybody has to have exactly the same opportunity before new benefits could be offered.

FROM THE FLOOR (unidentified speaker): I'm not so much worried about the fairness issue, because all of us are constantly trying to improve ourselves or our children. Even poor parents might try to send their kids, say, to piano lessons by saving up. But what does concern me in what you've been proposing are the risks. After all, we usually don't think it's okay to give our children steroids to improve their athletic performance, because of the side effects. So when we're talking about the risks associated with enhancing, are they worth it? Should we even allow them? Where do we draw the line? If somebody has a disorder, enhancement may be worth certain risks. But that's not necessarily true when we're talking about, say, privileged kids.

CAPLAN: Think of the history of growth hormone and kids. It started off by treating diseases, then it crept over to kids who were just a little short. And it was used in boys, not in girls; the culture says that short men are not as acceptable, normatively, as short women. So here was a risky drug that parents did use simply to get a little more growth when the "condition" clearly wasn't dysfunctional, just undesirable in some social sense. We could have tried to reform "heightism" in society and say, "Hey, get over it," but we didn't. We instead had pediatricians prescribing something that carries some risk.

I draw two conclusions from that. One is that we need tighter controls on risk, particularly to protect those who can't pick. Parents will push their kids, and that's dangerous; the kids are not choosing—they're being manipulated.

The other conclusion is that in doing our jobs as academics we should try to push on those norms and standards and boundaries that people have—the ones in place about beauty and performance and so forth are not beyond critique. So, for example, I don't actually think the right answer to shortness of height is growth hormone. It's in being more comfortable with who you are.

RONALD DE SOUZA (University of Toronto): The most powerful of the arguments you mentioned is in fact the stupidest—namely, that it's not natural, as if God gave us clothes, cars, and planes. As a lady said in a

New Yorker cartoon, "Nature intended for us to drink while flying." So my question is, How are you going to address the rhetorical packaging question? Tell me about how you're going to package your proposal.

Conference attendee Dr. Ronald de Souza, (University of Toronto): "The most powerful of the arguments is, in fact, the stupidest."

CAPLAN: One way to do it is by making frequent appearances on daytime television, sending the message out to those God-fearing folks. A more basic way is through education. One thing I found disappointing about the ELSI project was its failure to engage high school kids in discussions about bioethical issues in genetics. Neuroscience would be well-served to learn a lesson on that one and start young, and not because the religious view is wrong. It's because getting people to think about these things early, it seems to me, is the way to build an "adaptation" to—a familiarity and comfort with—what science has to offer. You can then intelligently choose an option, or not choose it, as you wish.

Not far from where I live are the Amish, and they don't really want to improve, much less familiarize themselves with the science-based tools I'm talking about. They believe that every child is a gift from God and they don't bring their kids in for treatment, even when they have genetic anomalies and diseases. It's a tough job to sort of push them and say, "Your child doesn't have to be that way," when they're saying, "Well, that is the way it has to be." I respect their coherent, if you will, value outlook; but at the same time I want to reach through to the next generation of Amish and say, "Think about this. You might want to modify that or move it in certain directions."

SAFIRE: Aren't you worried that by giving a pill or implanting some device in the skull we bring an unwanted equality to humankind—that there'd be no real differences in opinions or intelligence?

Human beings, at the end of the day, are going to pick capacities and abilities across a wide spectrum.

CAPLAN: This kind of scenario comes up when people ask about picking physical appearance. The bigoted among us might think we would pick only blue-eyed, blond-haired, tall, pectorally enhanced people. Actually, I think there'd be a lot of diversity. Human beings, at the end of the day, are going to pick capacities and abilities across a wide spectrum. And that's just fine with me, because I don't want to make our kids and their descendants all the same, but I do want them to have more choices. Not homogeneity, just more opportunity.

ELIZABETH WEISE (*USA Today*): The presumption seems to be that if you change the ability to remember things or if you change some form of intelligence, all other things will remain equal. But I have to say, as a journalist who covers science and technology, that very, very intelligent people are often a little odd. I don't know if it's Asperger's or *what* it is, but there's a certain constellation of psychological effects that one finds in extremely smart people. So if you start to change certain things about the brain, I'd expect that other things will change too. So the question for me is not "Is this right or wrong?" but "What is the law of unintended consequences?" What else are we going to find in these people who have in some way been affected? We probably have no idea what that is, which presents an even larger question.

SAFIRE: Good question; now answer that one.

CAPLAN: Well, that's what you do when you parent. You have to make decisions about how you'll try to shape and manipulate, environmentally and culturally, the next generation. It's risky and it often comes out wrong, and I would be the first to admit that I'm not sure we know exactly what we're doing in this enterprise. Nine times out of ten, though, it doesn't really matter; the kid is driven along by a far more complicated set of forces. Still, I do believe in unintended consequences, so we do have to monitor. Similarly, it would be wrong to engage in these technologies without careful, if you will, oversight and follow-up. For example, I'm a person who has yelled, for about two decades now, for some kind of monitoring of in vitro fertilization babies. If you're going to do it, you owe it to these kids to make sure you're not hurting them.

ROBERT LEE HOTZ (*Los Angeles Times*): I suppose this is a follow-up question. I wonder, given that we're talking about cognitive abilities and enhancements in otherwise healthy people, what sorts of novel challenges this kind of work poses for ensuring proper protection of those people who volunteer for such experiments.

CAPLAN: That's a good question, too. My colleague Paul Wolpe has come across this in some of the things he's done at NASA. When bioethicists make up a model for human experimentation, we usually presume that the subject will be there saying, "Well, if it's very risky, I'm not sure I want to do that." If you're an astronaut or a wanna-be astronaut, you say, "That's not risky *enough;* I'd like to do more. Could you centrifuge me another day?" So, too, would I expect that some of the pioneers in the world of, let's call it intellectual enhancement or mood enhancement or emotional improvement, are not going to be risk averse either; they may very well be risk *seekers.*

This gets us back to a basic issue of bioethics: Are we going to exercise more paternalism here? I think, in the interest of safety, that we should.

SAFIRE: Well, we promised we would discombobulate you at lunch, and we have. Thank you very much, Dr. Caplan.

Open Floor Discussion

ELLEN CLAYTON (Vanderbilt University): I have a couple of comments related to the earlier talks. One is that I think it's very important for this group to look squarely at the way the legal system will and will not use neuroscience information. We have to recognize, as Bill Winslade very appropriately pointed out, that while we can't undo the adversary system, we *can* elaborate some pretty straightforward principles about when the science gets good enough to be introduced into the legal system. After all, we've been down this road before, with lie detector tests and numerous other kinds of technologies, so it's critically important that many of us get involved and be very clear about saying what the scientific principles ought to be.

Children, and particularly boys, diagnosed with ADHD get put on Ritalin rather than kicked out of schools.

We also have to acknowledge the power of medicalization, particularly since it tends to create a more favorable social response and even some degree of entitlement to remediation. Children, and particularly boys, diagnosed with ADHD get put on Ritalin rather than kicked out of schools. I suspect that the pressure to medicalize will be even greater in neurobehavior than in other areas of biomedicine.

SCHACTER: It's interesting to me that while lie detection must contend with people's attempts to deceive it, in our memory studies they are doing the best they can. Yet we still have difficulty sometimes in distinguishing the true from the false. It's also interesting that different kinds of evidence carry different weight with people. On the one hand, evidence from lie detection is not admissible in court, yet when memory-related imaging findings are published there is this incredible excitement about them. People see that picture and think they're looking directly at the truth, when in fact the results reflect so many assumptions between the experimental design and the final image. So it's very important to keep these things in mind and to be cautious in the use of such "evidence."

WINSLADE: I agree. The burden is going to fall on the neuroscience community to carefully delineate what is and isn't known. Otherwise, there's a danger that things will be introduced prematurely and misused.

I remember an unexpected consequence of a PET scan experiment at UCLA in the early 1980s. The scans of a group of patients with Huntington's disease were compared with those of people at *risk* for that disease but who were normal. The researcher was startled to see that the brain images of both groups were indistinguishable. That kind of information is very important to have if such evidence is presented in a legal context; it helps guard against misinterpretation.

ARTHUR CAPLAN (University of Pennsylvania): This question is to Professor Greely, but I want to preface it by asking all foundation representatives and government funders to block their ears. I wonder what you think about the worthiness of the ELSI program? Probably the major achievements in genetics-related ethics, law, and social policy might arguably have happened anyway—not as a result of, but maybe in tandem with the ELSI project. So what should we learn about public and private funding from ELSI that can be applied to the nascent field of neuroethics?

GREELY: Great question. The ELSI program certainly

has provided lots of money for philosophers, anthropologists, lawyers, and others, and as a lawyer I think that's a good thing. But the greatest benefit of the ELSI program is that it has successfully provided political cover for the Human Genome Project, which is quite an important scientific accomplishment. The ELSI program has done very little harm, and that's good. And it has done some good in pushing faster some things we probably would have done anyway. But to segue into the second part of your question, I do not think it has done as much good as it might have, in part because there are some inherent constraints on what government-funded ELSI-type programs can do. They are limited in the issues they can consider and the things they can say.

I know there was tension within various ELSI bodies for a long time over how much policy work they could do: Should Congress be appropriating money for somebody who ultimately will say it should be spent on this or that? The lesson I would draw is that while it is useful to have an infusion of government money to study these issues, it would also be really important to have nongovernment money. This can bring different perspectives, and the more perspectives, the better. Also, private money can look at things that public money is less likely to look into.

IRA SHOULSON (University of Rochester): With regard to memory therapies, I was interested in Dan Schacter's comments about the potential ethical distinctions between treating an impairment such as Alzheimer's disease and doing interventions that enhance memory in normal people. And it strikes me though, that, both of these are going to be developed. I remember ten years ago when we first had botox to treat dystonia, we argued about whether to use it on people with idiopathic dystonia or people with blepharospasm. Well, look where we are right now. So I think the market will ensure that those interventions are there. And then the issue is, to what extent, if any, should they be policed?

Another point, about the implications of prediction that Hank Greely and others mentioned: maybe we're overblowing this a little bit. If you look at Huntington's disease, for example, only 3 percent of the people at risk for it actually get tested.

My last point is about the ELSI program, which [Professor Greely] brought up. The Human Genome Project is spending 5 percent of its funds, as mandated, to look at ethical, legal, social implications. In the area of neuroscience, at least as far as earmarked funds are concerned, 0 percent is being expended on neuroethics. What proportion *should* it be? And what kind of mechanism should be in place for funding neuroethics?

SCHACTER: My concerns about the memory-drug issue are twofold—one that's pragmatic, and the other that's more theoretical. The first is for an awareness of some of the sticky questions that would arise if these substances were to become available. Think of one's own child: all her classmates may be taking the new and effective ginkgo, or whatever, and let's say it really works. She could lose fifty SAT points if she doesn't take it too. Those are tough kinds of questions, I think, which I think people have to grapple with.

A more theoretically based concern comes from

my own analysis of the seven sins of memory. One could always look at these memory imperfections as flaws in the system—evolution fouled up a little bit, with deficits that only get worse when you have Alzheimer's disease—and they need to be corrected. But my own view is different. My argument is that you can look at a number of these "failings" as the flipside of adapted features of memory. Take the last one—persistence. After a traumatic event, it's troubling to be kept up at night by intrusive memories. But on the other hand, we've got that emotional memory system allowing us to remember threatening events so that we don't go there again.

After a traumatic event, it's troubling to be kept up at night by intrusive memories. But on the other hand, we've got that emotional memory system allowing us to remember threatening events so that we don't go there again.

So we should question what it would mean to get in the way of those normal functions, which I don't see as flaws in the system. What would be the costs of interfering with these normal aspects of memory? I don't have an answer, but I think it's a worthy question.

GREELY: I'll take questions number two and three. Yes, you're right about Huntington's disease. People at risk don't get tested for it that often, but this just points up some of the complexities of genetic testing in general. How many people get tested depends on a lot of different issues, including whether there is good medical intervention or not. Essentially a hundred percent of newborns get tested for a genetic disease called phenylketonuria, for example, because there is a very effective intervention that prevents its worst ramifications. So with neuroscience predictive tests, as with genetic predictive tests, some will be more attractive than others, though each will pose its own distinctive set of problems.

On the funding question: First, I don't have any particular percentage for neuroethics in mind, but at this stage in the field I would think a smaller rather than a larger amount would be appropriate.

Second, drawing from the ELSI experience, I'd like to see grants and research projects on social implications tied more closely to scientific research projects—and maybe funded by the same group. In the ELSI program, things sometimes got disconnected from what actually was scientifically realistic or plausible. I would try to counteract that. I would have a small NIH intramural program to help provide guidance for the broader extramural program.

And then, most important, I'd try to make sure there were private programs as well as public ones to provide a different perspective—to ask questions that public money is not likely to look into very deeply.

WINSLADE: Can I ask Hank a question? You seem to separate public and private funding. What do you think about public-private collaborations?

GREELY: They have some of the virtues of each and some of the vices of each. They are their own entity, and I think they are often a good thing, though they have their own distinct set of disadvantages as well. I didn't mean to rule them out.

STEVEN HYMAN: Hank Greely's example of Huntington's disease is a fine one, but it has little relevance to the behaviors that most people worry about, such as aggressiveness or whether somebody's prone to being addicted. And as Hank noted later in his talk, there are two points to bear in mind. The first is that almost all behavior is conditioned partly by the interaction of many, many genes, plus the environment, plus something we generally don't like to think about but that is very important—*chance.* Given that the wiring up of some hundred trillion synapses can't happen identically even in identical twins, the prediction of even the most "genetic" of mental illnesses, such as autism, can only be probabilistic. The public doesn't understand that, nor do insurance companies or employers. This is a real problem, and we have to address it.

The other point, which Professor Greely implied, but which I would like to make explicit is that genes do not control behavior. It's the nervous system that controls behavior, and there are many steps between the genes and the nervous system. Art Caplan has argued for a long time that limiting the ELSI focus to genes only puts off the evil day for our field, because real predictive power will come not from studying the twenty alleles in aggregate that give you a 40 percent risk of manic-depressive illness or a 30 percent risk of being somewhat more aggressive than the next guy; it will come from studying the nervous system.

When I was at NIH, the reason I didn't want to have a set-aside for ELSI issues was that behavioral neuroscience is still in its early days. I think that a critical step *toward* a more robust field—one that could do

ELSI-type research, which is very important, given the level of public misunderstanding and the implications for behavior—is gatherings like this. At meetings and workshops we can ask ourselves hard questions, but we must also try not to get ahead of ourselves in terms of what our science or medicine can really do. And it's good to have many disciplines involved. Medicine, in its enthusiasm, tends to overpromise, so it's really important to have people call our bluff and get us to engage in a dialectic. We need many more discussions like this one before there will be enough sophisticated people out there to warrant anything like set-aside funds.

ANTONIO DAMASIO (University of Iowa): I just wanted to pick up on something that has come back several times in questions from the audience and also in comments from the panel. The issue is what we can expect from functional imaging—namely, from PET or fMRI. We have to distinguish very clearly the uses that are diagnostic—excellent in some cases—from those that are not.

For example, no one need have any doubt that we can now identify a lesion caused by a stroke, tumor, surgical incision, or head injury, and that we can localize it and intelligently combine that information with clinical data. This enables us to make very accurate diagnoses and even predictions about how the person is likely to evolve. And I don't have any problem with that being brought in court, if needed, because it may help justice. The information is very robust, and we can count on it.

Most of the imaging issues that people are very

worried about have to do with functional imaging in an experimental setting. Here the interpretation is tied to the hypothesis, to the design, and to the theory that are behind a given study and to analyses that vary from laboratory to laboratory. This is the kind of information that we have to be very cautious about and that I would not find appropriate to introduce in court at this point.

I think we have to be quite humble at present; we are, after all, at the beginning of the experimental use of this approach. But I would not be completely skeptical about the possibility that in the future we may have very substantial results that would be usable in court.

BARBARA KOENIG: Would the panel like to respond?

WINSLADE: I agree.

KOENIG: That was simple.

I wonder if there's a legal distinction between behavioral problems that have their root in a head injury and those that derive from something else.

NOAH FEINSTEIN (The Exploratorium): Dr. Winslade, you mentioned in your case study that John had an auto accident, which was the likely cause of his symptoms, and that the people who made the diagnosis of schizophrenia did so on the basis of John's symptoms. But I wonder if there's a legal distinction between behavioral problems that have their root in a head injury and those that derive from something else. Is there a precedent for distinguishing between those things? Do you think there should be? What is the possible nature of that distinction and what are its implications?

WINSLADE: Because John was so completely whacked-out, there was no purpose in even having a trial. But the problem is that nobody thought about what to do with him afterward. One of the consequences of the simple "not guilty by reason of insanity" agreement between the prosecutors and the defense was that he went to a psychiatric hospital. He was treated there in ways that were in good faith, I think, but it became clear fairly early to psychiatrists that he wasn't being appropriately treated for the things that he needed. More precision in diagnosis, with clinical techniques that were available at the time, could have helped in the long-term management of his care.

MICHAEL WILLIAMS (Johns Hopkins University): I'm a neurological intensive care specialist, and hearing some of the comments this morning, I wonder why we keep talking as if there are *other* persons who have to have better understanding, better skills, and so forth.

In end-of-life care and some of the broader areas of ethics, ultimately it comes down to interactions between health professionals and individual patients or their families. If we don't provide these health professionals with the skills to analyze issues on a case-by-case basis, bringing to them an understanding of science and the public policy, we'll continue to have problems. I see as much misunderstanding among physicians who refer patients to me as I do among patients who've done their own research on the Internet. So we must not overlook the opportunity here to define the scope of training and teaching competencies in neuroethics—how to best impart these skills to the people who go out and practice and do research in the neurosciences.

GREELY: I agree with you entirely, although I just want to emphasize that neuroethics, if it becomes a specialty, is not just an issue for health care providers and about health care that's being provided. It's equally an issue for legislators, insurance companies, employers, school districts—for a wide range of human endeavors. The medical part is important, but we should not let it hog the limelight.

WINSLADE: I'll just add that it's very important to have collaboration and communication among different types of professionals, both for teaching students and for providing information to the public. Your comment about the Internet is an astute one; some people think that just because they can find it on the Internet, it must be credible!

DICK TSIEN (Stanford University): I represent the Stanford Brain Research Center, one of the sponsoring organizations for the efforts of Barbara [Koenig] and Judy [Illes]. I'm a basic neuroscientist, and I found the dialogue between Hank Greely and Steve Hyman to be fascinating because, in a way, they were both right and I like how they took the adversarial role—it sort of woke me up. It reminded me of the fact that neuroscience is a whole continuum, from the monogenic neurological diseases that Hank focused on to the more complicated things, like drug addiction, that Steve represents.

And what I took from Hank's talk about ELSI was the advice of an older brother: we've been there before, we've gone through all of these types of issues; so in launching the field of neuroethics, don't think you're figuring everything out for the first time—many of the same issues have arisen in genetics. That discussion also crystallized what, for me, is nascent support for the idea of neuroethics as a separate discipline. It made me realize the number of ways in which neuroscience is fundamentally different from genetics, even though it shares intellectual roots in the kinds of problems it grapples with. Fundamentally, the human brain differs far more from the *Drosophila* brain than the human genome differs from the *Drosophila* genome.

So my question for Hank is this: What concrete advice do you have to give us about where the geneticists were ten or fifteen years ago? Most of your comments were about the foolish pronouncements of famous people like Watson and Gilbert. What were the sensible things that people said at that time? Maybe someone like Paul Berg made statements that really changed the course of the field. And what sorts of advice should we be looking for right now?

GREELY: Actually, I've always wanted to be avuncular rather than an older brother. I've aspired to "avuncu-lardom."

Lots of people have said for a long time that many of these issues are quite complicated, that it's important for us to educate the public to understand these complexities, that problems aren't usually caused by single-gene, highly penetrant disorders. That has been counterpoised by short-run advantages—for some scientists, for the press, for others—in focusing on the dramatic. One lesson is that you cannot tell people too often that things are really complicated and not as dramatic as they think, because people want dramatic stories. This is a very strong force. Neuroscientists will have to fight against it and not be tempted. In particular cases it could be very easy to go with the flow, tell the dramatic story, and leave out the footnotes. In the long run, though, I think that having the public educated as to the complexity of these issues is our best hope.

Another useful thing that some people—Paul Berg certainly among them—said early on was that it is crucial for the scientists to remember that they work within the society. There is no Constitutional right to get paid to do whatever research you want to do, and ultimately the public will—and I would argue *should*—have some control over your direction and scope. In a sense it is like informed consent; one of the things that informed consent does is remind doctors that they really work for the patients. Similarly, it is useful for scientists to remember that they really work for the society in the long run and that they need to bring it along.

It is useful for scientists to remember that they really work for the society in the long run and that they need to bring it along.

Scientists also need to be humble in that regard and not think it is just a matter of educating people who don't understand what *we* understand—that if they knew the science as we know the science, they would agree with us and everything would be fine. In fact, it is a mutual education. Scientists will learn from people who know things about society and politics and cultures that you *don't* understand. So I guess I'd sum it up as humility. Learning humility is always a good thing.

KOENIG: I'd like to exercise the chair's prerogative to briefly respond to that question too, as another person who has spent years and years thinking about some of these genetics issues. And I'd like to point out that public education, and interdisciplinary efforts in general, are hard. We're dealing with people who have completely different ways of understanding and seeing the world and formulating problems.

And this leads me to another point, which is that one of the shortcomings in the way science is done is that it tends to leave out the social, largely because we don't have a lot of mechanisms today for how to study those kinds of interactions. Some people suggest that we simply leave it to "the market," but my work on the regulation of new genetic tests leads me to believe that throwing these things out into the market is not necessarily a way to solve problems.

ALBERT JONSEN: To pick up on Dr. Tsien's question about whether anybody said anything good in the old days relative to ethics and genetics, we do tend to quote the dumb things that famous people have said, but in fact two very important positive steps were taken in the late seventies, early eighties.

One was a report by the Hastings Center, which had a study group focusing on the appropriateness of genetic testing for various indications. This was a time when there was a powerful and enthusiastic rush toward testing everybody for everything, and there had been a disaster involving testing for sickle-cell anemia. The Hastings Group, after a year or so of study, came up with very sensible recommendations that have in large part determined the course of testing ever since. Many questions still come up, but the recommendations set a pattern for the way in which one could evaluate appropriate conditions and circumstances to initiate tests of various sorts.

The second contribution was a report by the President's Commission for the Study of Ethical Problems in Medicine called "Testing and Screening for Genetic Disease." It looked very carefully at the kinds of testing being proposed, analyzed them, and also set a pattern much like that of the Hastings report. So in the early years, two team efforts examined specific problems—issues that were ready to go out of control had those sorts of conceptual efforts not been made.

One of the lessons we might learn from the experience in genetics is that when considering the promise and the threat of new technology, it's important to distinguish between principle and practicality.

COLIN BLAKEMORE (Oxford Centre for Cognitive Neuroscience): One of the lessons we might learn from the experience in genetics is that when considering the promise and the threat of new technology, it's important to distinguish between principle and practicality. I'm intrigued that the word eugenics hasn't yet been uttered at this meeting, and it's probably because this is a word that's so abhorrent to most of us that we don't want to confuse the argument by mentioning it. But I suspect that most people here would sign on to the principle that it's good to do what one can to improve the lives of people—and that means, in some circumstances, improve their genes.

What we'd object to is the problem of who decides what's appropriate and what methods to use for implementing it. Killing people in order to achieve eugenics is not acceptable, but maybe gene therapy or the pharmaceutical manipulation of a gene product is completely acceptable. The question of the acceptability of the *principle* of eugenics is very much with us again in the new context of practicalities. Equally, I think, a lot of the very negative reaction to the concept of human cloning, at least from scientists, has been based much more on one's knowledge of the practical difficulties in achieving it—judging from problematical results in other species—rather than on the concept itself.

Finally, if this meeting is partly about coining new words, could I coin the new one *euneurics?* It would reflect a lot of what we're thinking about—namely, do we have the capacity to make people's brains better? Here, too, I would posit that some of the horrified reactions to the techniques of the past were largely based on their practicalities. Sticking needles into people's orbits and chewing up their frontal lobes are unacceptable ways of making people with disturbed behavior more pacific. By contrast, giving them a pill to pop is acceptable. So I think it's important to distinguish between principles and practicalities as we look at the ethical issues.

GREELY: I'd like to comment on both of the last two speakers. Certainly it is important to distinguish between principles and practicality. It is also important not to—I hate to use this word as a verb, but I can't think of a good alternative right now—privilege one over the other. The principles are important, but so are the practicalities. And, picking up on Al Jonsen's comment, focusing on discrete, near-term issues—problems of either principle or practice but problems ripe for answers—is a very good way to go. The genetic testing reports that Al mentioned had a good impact because they looked at near-term issues that were ready for consideration and they looked both at the principle level and at the practical level of who's going to do what, to whom, and why. *Eugenics* presents another example. I did not use the term *eugenics* in my talk. I talked about state-compelled selection, parental selection, and state-encouraged selection. I did that because *eugenics* has become so broad in meaning that it is not clear what someone using it means, other than that the thing described is morally bad. Being specific about what you mean by *eugenics* is crucial. Focusing on specific questions certainly isn't the only way to go—there is lots of value in broader and more principle-oriented discussions as well—but I think that one very good way to move forward is to focus on specific questions related to what is going to be happening soon.

Session

III

Ethics and the Practice of Brain Science

Bernard Lo, Session Chair

Professor of Medicine and Director of the Program in Medical Ethics, University of California, San Francisco

Steven Hyman

Provost, Harvard University

Marilyn S. Albert

Professor of Psychiatry and Neurology, Harvard Medical School, and Director of the Gerontology Research Unit, Massachusetts General Hospital

Erik Parens

Associate for Philosophical Studies, The Hastings Center

Paul Root Wolpe

Senior Faculty Associate, Center for Bioethics, University of Pennsylvania

BERNARD LO: Ladies, gentlemen, and others, I'd like to welcome you to Session III. This session will examine the ethics and the practice of brain science, and we have a distinguished panel. The first speaker will be Dr. Steven Hyman, who is provost of Harvard University and was formerly director of the NIMH [National Institute of Mental Health]. His talk is on the ethics of research and practice of brain science, and he will focus on psychopharmacology.

Dr. Bernard Lo, University of California, San Francisco.

Our second speaker will be Dr.

Marilyn Albert. She is professor of psychiatry and neurology at the Harvard Medical School, and she's director of the Gerontology Research Unit at Mass General Hospital. Her talk will be on ethical challenges in Alzheimer's disease.

Third will be Erik Parens, who is an associate for philosophical studies at the Hastings Center in New York. He will address the question, How far will enhancement get us as we grapple with new ways to shape ourselves? And our final speaker will be Paul Root Wolpe, who is a senior faculty associate at the Center for Bioethics at the University of Pennsylvania. His talk will be on neurotechnology, cyborgs, and the sense of self.

Ethical Issues in Psychopharmacology: Research and Practice

SUMMARY: Dr. Hyman pointed out the general efficacy and safety of psychotropic drugs, but he noted that while their immediate benefits are well understood, we really know very little about their long-term effects on the brain. Still, he was concerned that the nonuse of a drug, particularly for a child in great need, could have long-term impacts on the child's life at least as serious as any of that drug's potential side effects. Dr. Hyman explored the particular case of treating potential children with attention deficit/hyperactivity disorder with methylphenidate (Ritalin). This therapy is very effective, he said, but the problem is that too many kids with ADHD don't get it, and too many kids misdiagnosed with ADHD do get it. We need to understand the best use of existing treatments in all age groups, and we need better treatments, he said. And to approach that state, we need to do research—as long as the benefits outweigh the risks.

Dr. Steven Hyman, Harvard University.

STEVEN HYMAN: There are major ethical issues in psychopharmacology, both in research and practice, and in fact Art Caplan touched on some of them in the last hour. In my own talk, I want to focus not so much on issues related to treatment of illness or to enhancement but on the idea that drugs that immediately affect the brain and behavior may also have long-term effects. It goes beyond the early and simplistic model of psychopharmacology—you're depressed because you're down a quart of serotonin, so you get filled up and you feel better—to the more sophisticated idea that when you take psychotropic drugs, they may not only produce a short-term effect but also change the way your brain works, perhaps permanently.

Over the last several years, one of the really exciting advances in understanding the action of psychotropic drugs has been the recognition that when they bind receptors on the brain cell's membranes, they activate a set of biochemical processes that alter the cell's functioning. They signal, via complex biochemical networks, to the nucleus of the cell to turn on and off genes that are going to make protein products, which subsequently change the way nerve cells process information.

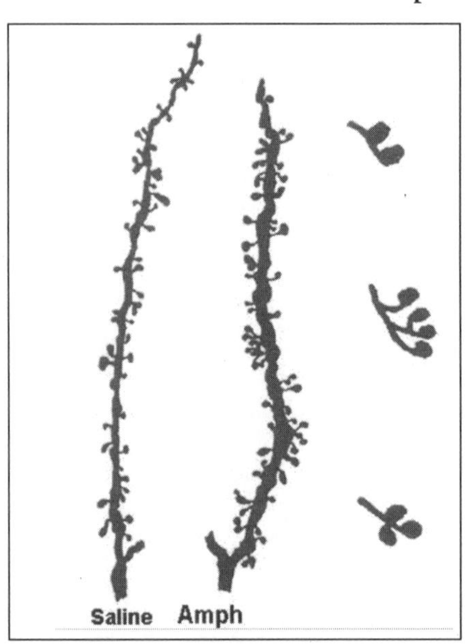

Psychotropic drugs change the structure of dendrites, remodeling the synapses and therefore changing the physical structure of the brain.

Specifically, we think that as a result of these signaling processes, psychotropic drugs change the structure of dendrites (the receiving surfaces of nerve cells) and, presumably, their synaptic connections. Consider these dendrites from two animals involved in an experiment by

Terry Robinson, one that got a saline injection and the other an amphetamine injection. The morphology of the dendritic spines, which in this kind of neuron is where a neurotransmitter, glutamate, is released, has really been changed by the amphetamine. Now, there's a lot we don't know—whether real connections are made, whether more glutamate is released—but these kinds of experiments do show that neurotransmitters and drugs can potentially lead to the remodeling of synapses and therefore to physical changes in the brain.

So how do such results and their possible permanence help us think—scientifically, practically, and ethically—about long-term effects? Consider addiction, which requires drugs, though it occurs on a background of genetic and environmental vulnerability. There is increasing evidence that the disease involves synaptic remodeling, and the resulting risk of relapse may last a lifetime. Other drugs—nonaddictive, therapeutic drugs—might similarly have very long-term effects on people, including children.

But before we really get ourselves into a lather, it should be noted that natural long-term memory also occurs by synaptic remodeling. So we have to ask how drugs differ from ordinary experience in physically changing the brain. And we have to ask something else. Because experience (recorded in different memory systems) also remodels the brain we must ask, How does having an *un*treated mental illness—or even a milder, untreated impairment—affect the developing brain or the adult brain? And how does it affect a person's life trajectory?

I want to take the most controversial example,

which is attention-deficit/hyperactivity disorder (ADHD). Kids who'd have been considered just bad kids in the past have now been medicalized with a diagnosis and a drug treatment, methylphenidate or Ritalin. Further, there's a concern that any hyper kids—boys and girls who are very active—may be given this diagnosis, which clearly is taking things too far. It's very important that ADHD not be diagnosed as soon as Johnny acts up in Mrs. Robinson's classroom. A diagnosis of ADHD requires inattention and impulsivity, with or without hyperactivity, at a certain level of severity and persistence. It has to occur in multiple settings, not just the classroom but also the playground and the home.

There is a diagnostic "gray zone" though, because this is still a clinical diagnosis. We do not have objective tests for ADHD. That in itself is very disappointing, because this is the kind of disorder for which cognitive neuroscience should be helping us find gold-standard laboratory-based tests. Unfortunately, they still don't exist.

We do not have objective tests for ADHD. That in itself is very disappointing.

When it's well diagnosed, ADHD is found in only 3, 4, maybe 5 percent of children, and it's diagnosed more often in boys than in girls because boys are more likely to be hyperactive and thus get attention. But one of the things about ADHD is that it's not just limited to childhood but extends to outcomes—often quite bad ones, actually—in adolescence and adulthood. It's associated with academic and occupational underachievement, higher risk of substance abuse, and even a greater likelihood of trouble with the law. It's also associated with heightened risk of other mental disorders: depression, anxiety, and conduct disorder.

Treatment

- Behavior therapy is effective
- Stimulant medications are effective
- Careful use of medication is more effective than behavior therapy (MTA 1999)
- Combining medications and behavior therapy is a valuable strategy for children with co-occurring disorders (MTA 1999)

What do we know about treatment? Stimulant medications like methylphenidate have been very well studied—actually, more than any other psychotropic drug in children—and they have been shown to be safe and effective for school-age kids with ADHD. (These drugs haven't been well studied in preschoolers, who are increasingly receiving them, though a study is under way.) One of the things that a very extensive trial on schoolkids has shown is that the appropriate use of medication—with the right dosage titration—is more effective than behavior therapy. And it also showed that you get very little additional benefit from combining medication with behavior therapy, unless the kid has a co-occurring disorder, in which case combined treatment is best.

We also know that community treatment is not as effective as it should be. Lots of kids get methylphenidate, but they don't have good outcomes presumably because health care providers do not educate families adequately, poor dosage titration, or inadequate attention to side effects. We also know that a very large number of children who receive stimulants do not meet the criteria for ADHD, and that a substantial number of children with ADHD are not diagnosed or treated in community settings. Generally speaking, kids with ADHD who don't get a diagnosis or treatment tend to be from the inner cities, and the kids without well-diagnosed ADHD who are treated anyway tend to live in the suburbs and are often middle-class or upper-middle-class.

There's also the issue of social pressure for medication. That is, families don't want to unilaterally disarm. If all the other kids are getting methylphenidate and sitting still and studying, why should Johnny be at a disadvantage? And besides, while there are side effects of methylphenidate, they tend to be relatively mild and manageable for most kids—though I'd point out that we really don't know its long-term effects on the brain.

Still, it's important that we not confuse problems in our health care system with the idea that there's something wrong with the medication or that there's something prima facie wrong with treating kids with psychopharmacologic agents. The bottom line is there's a mismatch between children in need of treatment for ADHD, and children getting the diagnosis. We need objective diagnostic measures. We need to understand the best use of existing treatments and we need more and better treatments.

To understand the best use of treatments in all age groups, we need to do research. And here there is substantial disagreement about the ethics of psychopharmacology research in children. (I received an unbelievable amount of hate mail about initiating a clinical trial, in children under the age of 6, to study the safety and efficacy of methylphenidate.) My view is that the drugs are being used more and more, and we don't know anything about safety or efficacy in young kids, so right now every kid is an uncontrolled experiment of one—and that situation is intolerable. Thus we need a carefully conducted clinical trial.

We also have to understand, as I mentioned earlier, that there may be risks of *no* treatment. If a kid has an untreated mental disorder and doesn't utilize school well and doesn't form normal peer relationships, there can be a downward spiral. It is not very easy to recover from eight years of poor school performance, peer rejection, and maybe an arrest or two. Such impacts may be at least as enduring as any hypothesized long-term effect of psychotropic agents.

It's very important to recognize that, as a society, we often treat psychotropic drugs differently from other drugs. Generally in medicine we talk about the need not just to treat illness but to prevent or to intervene early—which often means in children—except when it comes to psychotropic drugs. So we have to ask ourselves if there's a moral or ethical difference between altering neurotransmitter levels and, say, lowering cholesterol levels. Though it's deemed okay for large populations to be on a statin to lower cholesterol, that's not necessarily so for an SSRI (selective serotonin reuptake inhibitor) or a stimulant or modafinil—the drug that keeps you alert and engaged so that you can fly your B-2 bomber or listen to a lecture despite sleep deprivation.

It's very important to recognize that, as a society, we often treat psychotropic drugs differently from other drugs.

Ethical issues in Psychopharmacology Research and Practice

- Psychotropic drugs have immediate effects on brain and behavior
- They also have persistent effect

As researchers and health care professionals, we have to address such issues. NIMH's attitude about a clinical trial in preschoolers for Ritalin was not to say that we couldn't use drugs in children this young but to recognize that these kids were in a risky state and we were going to find out whether the treatments were safe and effective. And we asked, Does the research warrant the risk? That is, did we really need this information? Well, we thought that if 1 percent of preschoolers are already on this drug and there's a lot of evidence that it's safe and effective in kids over the age of 6, the risk-benefit ratio—what societies need to know versus the risk to these kids—ultimately justified the research.

Can we design a study to minimize risk without compromising the value of doing the study in the first place? This is often a very controversial idea. The design here was that the kids were very severe—the sort who've been kicked out of multiple day cares—and they had to undergo intensive behavioral therapy before they were randomized to a drug arm (if, after some eight to ten weeks, they failed behavioral therapy).

Can we achieve appropriate informed consent? In general, whether for kids or adults, I think it's very important to recognize that signing a form is a *receipt* for informed consent; it's not informed consent per se. Informed consent is an ongoing educational process, and in this case it obviously means involvement of the family, and more than that—for example, the parents get reconsented on a regular basis.

So I think we *can* do this kind of research if we grapple carefully with such issues. I think that we as a community, and as a society, have a set of complex and not fully resolved ideas about psychopharmacology—in all patients, but especially in children—that is a very rich topic for continuing discussion.

Ethical Challenges in Alzheimer's Disease

SUMMARY: Dr. Albert discussed some of the ethical issues at each of the three basic stages—presymptomatic, preclinical, and actual clinical dementia—of Alzheimer's disease. In the presymptomatic stage, she noted, a negative genetic test result could be a false negative because many of the possible causative gene mutations are not yet known. She also pointed out the absence of confidentiality; although the results of the test may be confidential, the fact that the test was done goes into the patient's medical record. For the preclinical stage, Dr. Albert observed that as prediction methods improve and treatments (not necessarily benign) are developed, patients must be educated in the concepts of probability. This is because test results are likely to only indicate risk and not offer a deterministic outcome; there are likely to be multiple gene interactions, and all may be affected by environmental factors. For the clinical dementia stage, she discussed issues of who might provide informed consent for the patient, and under what circumstances; this is a serious concern even for patients who appear to be only mildly impaired.

MARILYN S. ALBERT: When I arrived this morning I ran into a colleague, and I told him I'd be talking about ethical issues in Alzheimer's disease. "Well," he asked, "are you going to come out against it?"

He was joking, of course, because everyone is indeed "against" any debilitating illness. But we're all against Alzheimer's in an even more profound way: it is one of the diseases that we all fear the most because we are creatures who use our brain and identify so deeply with it. Not only would we lose our intellectual capacities with Alzheimer's disease, we'd lose our individuality—who we are as people.

It's also a disorder that represents many of the issues we've been talking about at this meeting—it spans the range of ethical choices and challenges that we have before us. So I'm going to talk about what these ethical challenges are, and I'll relate them in particular to the different stages of Alzheimer's disease; it's very clear that the questions a patient (or patient's representative) has to face, and that we should talk about here, depend on the stage of the disease he or she is in. The three basic stages are presymptomatic, preclinical, and actual clinical dementia.

Right now, the only way to know for sure that someone has Alzheimer's disease is to examine brain tissue. That's normally done after a dementia patient has died, and we look for certain pathological abnormalities that we call neuritic plaques and neurofibrillary tangles. But in actual clinical practice in most places, the diagnosis is made according to criteria that were established in 1984 by Guy McKhann and his colleagues. Following these criteria provides a high degree of accuracy—about 90 percent.

Dr. Marilyn Albert, Harvard Medical School.

Three treatments have recently

become available on the market, though they are exceedingly modest in their effect. On average, they produce about a six-month improvement in behavior, but they don't slow the course of the disease. Their big benefit at the moment is that they have very few side effects. Yet the absence of side effects is one of the things that I believe has been leading people to largely avoid thinking about what the future ethical challenges regarding early diagnosis and treatment of Alzheimer's disease may be.

Let me begin by discussing some of the ethical challenges with respect to presymptomatic treatment—the first stage of Alzheimer's disease. They involve people with dominant genetic disorders very much like what was mentioned with respect to Huntington's disease. As it happens, there are three dominant genes in Alzheimer's disease. If mutations occur in any of these genes, the individual will definitely get the disorder. Complicating the picture for AD is that the gene that is most commonly mutated (among the three) has over eighty different mutations that have been identified so far.

This means that if you talk to people about whether or not to have genetic testing, they must recognize that the test may not find any of these mutations; many of them are what we call 'private'—unique to a family. It would be time consuming and very costly to look beyond the known mutations to see if, in fact, the test has revealed a new one. But if negative results were certain, there are other implications of the testing.

I participated in the planning of the genetic counseling group at Massachusetts General Hospital, and one of the things we discovered is that if you just send a test to Athena Pharmaceuticals for evaluation, the fact that a blood sample was sent for genetic analysis ends up in the patient's medical record. For that reason, all the consent forms we have at Mass General—having to do with genetic testing of any kind—specify that the hospital cannot completely guarantee the confidentiality of the information. Someone might not be able to find out the results of the genetic test but they might be able to find that a genetic test was done (either for clinical or research purposes). So this has to be clear right from the outset.

All the consent forms we have at Mass General—having to do with genetic testing of any kind—specify that the hospital cannot completely guarantee the confidentiality of the information.

This is one reason that, as in the situation with Huntington's disease mentioned earlier, we're anticipating that there won't be a great call for such testing. In addition, of course, people at risk face many different choices: whether or not to get married, whether or not to have children, the impact of genetic testing on insurance or employment. We need to have further discussions on why in fact so few people go out and get tested, though obviously we can offer some good guesses.

Now let's consider the other end of the spectrum, the stage of clinical dementia. First, with respect to people who have substantial cognitive impairments, we make the assumption that these individuals can't understand risk or future planning, which has clear implications for their capacity to provide consent for clinical care or research. It's obvious why we all assume this, because that capacity would have to include many different cognitive skills, including language, executive function (which involves self-monitoring, planning, and understanding the future implications of choices), and memory.

With respect to clinical practice, the current standards for significantly impaired people are fairly well defined. The next of kin can give consent for extraordinary procedures. But while extraordinary procedures are best served by guardianship, it is very hard to get; you have to go to court, you have to have someone evaluated, you have to have a physician's evaluation that the patient is not competent—and even then the case often needs to be adjudicated.

With respect to consent for clinical research, there is not a lot of agreement about how people who are moderately or severely impaired should be treated. There are federal guidelines, but very little agreement from institution to institution. So it's really left up to individual IRBs (institutional review boards) to make the decision, and because the composition of these IRBs varies, their skill in these matters varies as well. Most important, there is not very much agreement about who can serve as a surrogate in most instances. Federal law permits what's called a legally authorized representative, but it's not clear who that person might be. If there's a legal guardian, then he or she is obviously acceptable. But if not, there's often a problem: the next of kin may not be available. Who a good surrogate may be in such cases is another issue we need to discuss.

Federal law permits what's called a legally authorized representative, but it's not clear who that person might be.

Things are even less clear when it comes to Alzheimer's disease patients who are mildly demented. We don't even talk about this very much because people who are mildly demented appear, superficially, to be quite normal. And that leads us to believe they can make a lot of decisions for themselves about their clinical care and whether or not they should participate in research. But recent research in fact tells us that a person who is mildly demented with Alzheimer's disease, apart from having a memory disorder, is also likely to have diminished executive function—thus, the patient may have limited insight into their disorder and have difficulty understanding the implications of decisions.

You might think that it's not so important to make a critical decision if someone is only mildly impaired, but some patients progress at a very rapid rate. Therefore we need to think now about what ought to be done for them with respect to informed consent. At present, proxy consent is often provided by next of kin, other family, and friends. And as Steve Hyman said, what typically happens is that the proxy doesn't just sign a consent form; they need to be involved all along the way. Both in research settings and with respect to clinical care, you want them agreeing to things even though that isn't something that anybody has established as legally binding.

The third and last phase of the disease I'll discuss is actually in between the first two; it's the preclinical—or prodromal—phase of the disease. And this is the phase about which there has been the least amount of thought and discussion. It's important for us to remedy that situation in light of the enthusiasm among Alzheimer's disease researchers regarding potential treatments. They predict that within a very short time—a number often mentioned is ten years—we will have truly effective treatments for Alzheimer's disease; the reason for this confidence is that scientists believe they now understand the basic mechanisms of its cause. And when those treatments come about, it's unlikely they'll be as benign as the current ones are. Issues of consent will loom ever more important.

As will testing and prediction. But although a great deal of work is going on—in our group and around the country—to ultimately enable us to predict who's going to develop the disease down the line, few have thought about the accompanying ethical challenges: What kinds of things should we say to people, for example, and what kinds of certainty should we express in saying them? Who among us would want to wait until we had lost a great many essential neurons in the brain before we were treated?

The one gene that has been identified with respect to late-onset Alzheimer's—the ApoE-4 allele of the ApoE gene—only carries risk for the disease. It isn't a deterministic gene.

You might think that by the time we got effective treatments we would also know a lot about genetics and be pretty certain about our predictions. But as you heard earlier today, the one gene that has been identified with respect to late-onset Alzheimer's—the ApoE-4 allele of the ApoE gene—only carries risk for the disease. It isn't a deterministic gene. In addition, genetic epidemiologists and statisticians have estimated that, based on what we know now, there are likely to be as many as five genes that affect late-onset risk. Some of these might increase risk, while others might be protective. But all of them are likely to interact with each other and be affected by environmental factors.

So in essence we'll be trying to identify people early and tell them something about their risk, and if current research fulfills its promise, we'll be able to inform them at earlier and earlier stages. They, in turn, will have to learn probability theory in order to deal with the information. One of the really important areas of investigation with respect to the ethical and legal challenges of Alzheimer's disease, I believe, will have to do with educating people on risk and concepts of probability, and we should begin discussing it now.

This issue of probability is relevant not only to our thinking about the brain and brain diseases but to lots of other disorders as well. Diseases of the heart, breast, and lungs, for example, will likely turn out to be affected by multiple genetic problems as well. In fact, I must mention that many of the things that I've been talking about with respect to Alzheimer's disease also apply to numerous other conditions, especially on issues that have to do with cognitive incompetence.

To move forward on the ethical issues I've discussed and many others, there has to be increased regulation and legislative input. We need to have IRBs that are better informed. And what especially needs to be done is to invite advocacy groups around the country to play a bigger role. If, for example, the Alzheimer's Association takes a particular stand with respect to the ethical and legal issues regarding prodromal disease and prediction of disease, that could greatly influence legislation and the decisions of IRBs.

How Far Will the Treatment/Enhancement Distinction Get Us as We Grapple with New Ways to Shape Our Selves?

SUMMARY: Dr. Parens suggested that the treatment/ enhancement distinction can be one tool among many that we employ as we contemplate how to use psychopharmacological agents to shape our selves. The distinction can be a place to begin—in deciding, say, what to include in health care coverage, or in affirming natural variation. In cases where the treatment/ enhancement distinction can't help us—as in already entrenched practices—Dr. Parens suggested that we "specify the consequences we are hoping for or fearful of" and recognize that different ways of dealing with an issue reflect different values. He offered several examples of phenomena with the (negative) consequences of unfairness, complicity with unjust norms, and homogenization.

ERIK PARENS: I recently encountered Francis Fukuyama's book *Our Posthuman Future: Consequences of the Biotechnology Revolution.* Though Fukuyama is on the opposite end of the political spectrum from Art Caplan, like Art, Fukuyama has an ear for a provocative thesis. Caplan says that neuroscientific advances are, ethically speaking, really just more of what we've been doing all along. Giving his kid a boost with a drug or brain chip or whatever really isn't ethically any different from sending his kid to a fancy private school. Fukuyama's thesis is that neuroscientific advances could threaten our humanity. His claim is that they may, if we're not careful, usher in a "posthuman future."

Fukuyama argues that to forestall that danger we need to understand and preserve human nature. And he asserts that medicine, being a repository of knowledge regarding not just health but so many other aspects of humanity, can play an important role in that quest. He says further that we need to establish a federal agency to try to distinguish between legitimate and illegitimate purposes, and he suggests that such an agency should rely heavily on the treatment/enhancement distinction (which contrasts the traditional goals of medicine—eliminating disease and restoring health—with the augmentation of otherwise-healthy individuals). I want to suggest that the distinction isn't as useful as Fukuyama hopes or as useless as Caplan suggests.

Erik Parens, The Hastings Center.

Many, many pages have been written regarding how blurry the line is between treatment and enhancement, and I'm not going to revisit that issue today. My question here is, What is the treatment/enhancement distinction useful for? It is a tool, after all, so what can we do with it?

Well, it's a place to begin. In articulating what goes into universal health care, for example, everybody can't have everything, so we might want to start distinguishing between things that go into the basic package of care and things that don't. Presumably the treatments go in, the enhancements don't. The tool can also be used to begin critiquing some social practices. If, say, shyness isn't a disease, perhaps medicine ought not to treat it and medicalize it.

Very closely related to that, the treatment/enhancement distinction can be used to help affirm natural variation. Should we be concerned that people who wield this tool may believe that treating diabetes or cancer is bad or unnatural? No, I don't really think so. I do think it can be a way of saying why it bugs us to give growth hormone to short kids who aren't growth-hormone deficient. That is, the treatment/enhancement distinction could help forward the argument that there are a lot of ways to be in the world—we come in all sorts of shapes and sizes—and we don't need to use medicine to change them. Again, it's a place to begin.

What won't the treatment/enhancement distinction help us with? For one thing, it won't help us articulate limits outside of medicine—which brings me to the "schmocter" problem. Imagine that in the not-too-distant future there are people who have access to new biomedical technologies but who don't call themselves doctors. They are self-declared schmocters who do not pretend to share the goals of medicine but rather think of themselves as pursuing the goals of schmedicine. To people who are practicing schmedicine, it's not going to matter very much to learn that their practice is inconsistent with the goals of medicine. It's just an irrelevant claim. In order to address schmedicine and its purposes, we've got to start making a different kind of argument, which I'll get to in just a second.

The treatment/enhancement distinction also isn't going to help us distinguish between enhancements that we've already begun to endorse and those that we don't want to endorse. Being capable of getting pregnant is surely not a disease. Giving people contraception isn't a treatment. According to the logic of the goals-of-medicine argument, giving people contraception so that they won't get pregnant is an enhancement—though one that most of us are incredibly grateful for. Similarly, menopause isn't a disease, and giving HRT isn't a treatment but a kind of enhancement that, again, is very widely appreciated and embraced.

Being capable of getting pregnant is surely not a disease. Giving people contraception isn't a treatment.

So if the treatment/enhancement distinction won't help us deal with these kinds of problems, what can we do? It seems to me that we have to specify the consequences we are hoping for or fearful of. Being in the ethics biz, I'll stick to the ones we're fearful of (though I'd remind you that with new technologies what we'd really like to do is less reactive and more positive—promote human flourishing).

Consider the consequence of unfairness, something we've talked about already at this meeting. In the example of ADHD and Ritalin that Steven Hyman so beautifully laid out, what would happen if anxious parents in the suburbs increasingly gave Ritalin to their kids—most of whom did not have ADHD—so that they performed better on the kinds of tests that give people access to the kinds of places that enable them to get still more advantages? It seems to me that it's legitimate to worry that the gap between those who have and those who don't might grow.

It seems to me that it's legitimate to worry that the gap between those who have and those who don't might grow.

Of course, as we've heard, the first line of argument is, So what? I mean, the rich have always had access to new technologies. So it seems to me that we're not talking about a brand-new problem here but the potential exacerbation of a very, very old problem. We're not just talking about kids with resources getting access to Stanley Kaplan. We're talking about kids with resources getting access to a capacity that will help them use Kaplan still better so that they can do still better.

What about if *everybody* had it? Would there be reason to worry then? Here it's important to appreciate that the means we use to achieve our ends really matter because they each express different values. Everybody agrees that improved performance is a wonderful thing; we all want it. But it seems to me that it makes a difference whether you change the teacher-student ratio to improve the performance or you give the kid a pill. In the first case you're expressing your commitment to the value of engagement. You want kids and teachers to interact better so that kids will do better. The pill, meanwhile, expresses the value of efficiency—which, God knows, is important.

But let's at least be clear that in using the different means to achieve the desirable end, we're expressing different values.

Another consequence I'm worried about is becoming complicit with unjust norms. Imagine there's a color-blindness pill that could reduce discriminatory attitudes toward dark-skinned people. Everybody agrees that discrimination is hurtful to those who are discriminated against, and that they suffer as a result of it. So if people are suffering, why not give this color-blindness pill to the people who are making them suffer? The fear is that in giving the pill we become complicit with the unjust norm that a particular skin pigment is preferable. Whether you take the pill or don't take the pill, the unjust norm remains in place; nothing has changed.

Then the question is, Should anybody's well-being be sacrificed on the altar of social justice? You know, we have this pill that would reduce discrimination, but we're not going to use it because we're worried about social justice. I believe it's really, really important to be exceedingly cautious about suggesting that anybody's well-being ought to be sacrificed on any altar. But it also seems to me that it's very important to remember that our different means of achieving purposes—in this case, reduced discrimination—express different understandings of who we are. In this hypothetical case, we'd be using the pill to change our attitudes—and we'd be expressing an understanding of ourselves as "mechanisms." This is different from thinking of ourselves as reason givers who can convince each other that discrimination is unjust. Same end, different means, and they express altogether different understandings of what it means to be a human being.

Another consequence of concern is homogenization, as in the example of Prozac. So what if more of us were more assertive and confident and resilient? But again, it seems reasonable to worry about this tendency to make more and more of us more alike. Peter Kramer, in his very interesting book *Listening to Prozac,* argues that Prozac is an all-purpose means. In a recent *Hastings Center Report* article, he essentially says, "Listen, don't worry about Prozac producing conformity. Prozac doesn't produce anything; it's good for any life project. It can be used for projects of conformity just as surely as it could be used for projects of rebellion or resistance."

We have to ask, as we survey any new technology: How will it actually be used?

It seems to me that his point, in principle, is absolutely correct but in practice altogether implausible. We have to ask, as we survey any new technology: How will it actually be used? Do we have reason to believe, in particular, that it might be used to exaggerate problems we already have?

So in answer to my initial question—"How far will the treatment/enhancement distinction get us?"—I think it's a place to start. I don't think it's going to get us as far as Fukuyama hopes. Appealing to "nature" isn't irrelevant, but it can't stand in for specifying the consequences that we hope for and fear.

Neurotechnology, Cyborgs, and the Sense of Self

SUMMARY: Dr. Wolpe celebrated the potential of "bionic" technologies—throughout the human body but especially in the brain—to extend our abilities and our lives, but he also urged caution. He cited a wide variety of "physiotechnologies"—a few already in place but most on the way (likely sooner than later)—"to be incorporated into our very flesh and become part of who we are." Some of these technologies, he noted, such as neuronal chips, could be new applications of what is already part of our flesh. Thus "we are technologizing the organic world and we're organicizing the technological world, and these innovations are going to have a profound impact on the way we live." Not everyone is looking forward to these changes, Dr. Wolpe observed, and in any case we all need to ask ourselves and each other some serious questions about how we wish to direct our own evolution. So we must pose these questions not only systemically but early—to help direct science and avoid being led by it or having to belatedly chase it.

PAUL ROOT WOLPE: Art [Caplan] and Erik [Parens] talked about the idea of posthumanism. But it wasn't the people they mentioned (Leon Kass, Francis Fukuyama) who came up with that concept. It has been pushed primarily by those who are for it—Ray

Dr. Paul Root Wolpe, University of Pennsylvania.

Kurzweil, Greg Stock, and others—who look *forward* to the moment we're posthuman or transhuman.

For these people, control of our own evolution is the ultimate goal. In their conception, the first stage of evolution was unconscious and physical; the second stage was a cultural intervention on that physical evolution (through the ways in which culture—medicine and other intellectual influences—has modified natural selection); and now we have a fusion of those two things, in which culture is going to determine our very physiological forms. In other words, human beings have taken over evolution, and this is going to end with our own conscious and directive alteration of our physiology. To the posthumanists, that is a good thing.

We're already acting on that agenda: we're sculpting flesh; we're genetically altering animals in order to transplant tissue into human beings and not provoke an immune response; we're even planning to transplant genetically modified *organs* from one species to another.

I always do a revealing exercise with my undergraduates. I tell them: "The pig heart is the same size as the human heart, so what we're going to do is genetically alter these pigs so that their hearts won't be rejected by the human body. And then we'll have this herd of pigs, and when somebody needs a heart transplant, we'll take a pig and slaughter it, cut out its heart, and transplant it."

And the students' reaction usually is: "That's terrible. That's just a terrible thing to do to a pig." I then remind them about bacon and ham and all the other reasons we slaughter pigs, for purposes that aren't life-saving, in order to illustrate this fundamental moral disconnect we have. I do that exercise with every class, and every class reacts the same way.

This is a very roundabout attack, or at least a dig, at Leon Kass's whole idea of the wisdom of moral repugnance. It is exactly moral repugnance that sometimes is misguided. The false notion that this sort of original visceral reaction is the right one is perfectly illustrated by the fact that the same people who are sitting there munching on their pig parts find some moral objection to saving someone's life with a pig.

So we really have to get beyond our initial ideas and try to understand the moral principles on which we're basing our decisions. We have to keep our minds open as the fruits of science and technology become more and more awesome yet at the same time routine. We're talking, for example, about taking nanotechnologies—microscopic computer chips, rotors, and other things—and sending them into our bloodstream to, say, Roto-Rooter out our arterial sclerosis. And some people are talking about flooding the brain with microchips to enhance certain kinds of brain processes.

What we're talking about is a whole series of "physiotechnologies"—that is, technologies to be incorporated into our very flesh and become part of who we are. We're already doing it, of course, with prosthetic limbs. Yet somehow we think of those limbs as separate from ourselves—unless you happen to use one. If you talk to people who have an artificial arm or leg that they depend on, and you ask them about their relationship to it, they'll tell you it's not just a piece of hardware that they strap on. They establish an emotional relationship with the object that becomes very much a part of their function.

Talk to people with the AbioCor artificial heart about their relationship to *that* piece of hardware. They'll probably tell you they hate it because it isn't fun to have this heart in you. On the other hand, they'll also tell you that they love it because it has kept them alive. And the innovations don't stop there. We now have artificial bladders, artificial lungs, even artificial kidneys if you want to so label dialysis. Maybe someday dialysis will get down to the size at which it can be implanted.

The pumps and bulbs and balloons of the body are only the first things that we can synthesize. Other things are coming, they're coming very quickly, and they're coming from all sides. So now we're talking about brain-implantation technologies, implanting fetal cells or computer chips. And they've been tested for a whole series of diseases so far—for Parkinson's, epilepsy, deafness, blindness, depression. The vagus nerve stimulator that was first developed to treat epilepsy—it's been called the pacemaker for the brain—has now been shown to possibly have an effect on depression.

We've learned to enhance the inherently slow perceptions of our brains.

This is no longer just science fiction but something we actually have the technological capability to attempt. We have learned how to hack into the wetware between our ears in the same way that we've learned how to hack into other information systems and change the patterns of communication, as we do, say, over the Internet. We've learned to enhance the inherently slow perceptions of our brains.

It's going the other way, too, with organic tissues serving technological functions. Scientists are using chip lithography to create "neuronal chips," silicon chips with furrows cut in them so that actual neurons grow in such a way that their dendrites and axons interact to create certain input and output patterns. Similarly, because DNA is the single best information store we know of, they're beginning to create DNA computers with storage capacities that far outstrip anything we can create synthetically.

Duke University's Dr. A. L. Nicolelis, the owl monkey, and the robotic arm. In a dramatic example of technology melding with the mind, the monkey moved the artificial arm from 600 miles away using brain signals transmitted over the Internet.

The technology is thus coming from both directions. We are technologizing the organic world and we're organicizing the technological world, and these innovations are going to have a profound impact on the way we live. With these kinds of new technologies, we won't ask the question, "Does the 'me' who can have a relationship with a prosthetic arm also have a relationship with this neuronal chip?" Unlike the limb, the latter technology will determine who that "me" is who's doing the appreciating. That is, once you integrate technologies into the brain, you then have to ask yourself the question, Is there an end of the non-technological me and the beginning of the technological me, or is it now *all* me? Am I part technology and part organic?

The point of all this is that we really are becoming some kind of cyborg, some kind of posthuman in the sense that for the first time in history we really are going to incorporate our synthetic technologies into the very physiology of our being—with major, though not necessarily entirely undesirable, consequences. Speaking as someone in the prodromal phase of old age, I think we can look forward to a time in the not-too-distant future when many of those symptoms or characteristics of growing old are going to be compensated for by these technologies.

Not everyone, though, is looking forward to it. Take a technology already being placed in the head—cochlear ear implants. There are people in the United States right now who say that cochlear ear implants deny them their rights as a subculture. They say that these kinds of devices are taking away something that is inherent in their self-definition and that they value—their inability to hear! So all these technologies are going to have to deal with opposition. Whether it's based on conceptions of "don't mess with human nature," subcultural values, religious values, or whatever, some people maintain that certain ways of being are correct and that technological manipulations of them are wrong.

We do, of course, need to ask broader questions

regarding how, as a species, we want to direct ourselves, and toward what goals, as we begin to incorporate our bionics into our bodies. And we can't ask the ethical questions once the science is already out of the barn. Rather than allow the science to pull us along, we must now begin a serious conversation—and not a conversation à la Francis Fukuyama, in which he postulates an inherent "human nature" that he never defines in a specific, historically and culturally consistent way. He thus then gets to call any particular trait he likes "human nature" without ever defending it.

Biotechnology is no longer a tool merely to try to bring the human body back to some preexisting baseline of function. We can now (or soon will be able to) push those boundaries, to "enhance" ourselves in ways we could before only imagine. Implantable computer chips are allowing the blind to see, the deaf to hear, and monkeys to control cursors on computer screens entirely with their minds. Transcranial magnetic stimulation can turn specific areas of the brain off temporarily by sending electric charges through the skull. Electrode implantation has allowed scientists to create "robo-rats," whose travels are controlled by the joysticks of scientists back in their labs, and monkeys whose thought processes can control mechanical arms thousands of miles away. Psychopharmaceuticals promise to increase memory retention, even out the rough spots in our moods, focus our attention when we do complex tasks. The posthumanists are right in their descriptions of the future, whether or not they are right in the enthusiasm with which they embrace it.

Ultimately, it will be the society as a whole that decides whether or not to embrace these technologies, and the pressure to do so—from the technophilic medical establishment, to the enormous biotechnological/ industrial complex, to the advertising departments of the companies that peddle the technologies—will be intense. We must very carefully project what the implications of so drastically reshaping ourselves will be. The way to do that is not to postulate an ill-defined "human nature" that we violate when we pursue these technologies, but to ask very practical questions about what our lives will be like, and how society will be shaped, if we employ each one. Some may be wonderful additions to our medical armamentarium, and others may have risks or dangers that are hard to foresee. But that is, I believe, the role of bioethicists and others who are trying to help guide us down the slippery and tortuous path of neuroethics.

Question and Answer

JOE DUMIT (MIT): I have a question for Steven Hyman about the effective overprescription of Ritalin in the suburbs.

STEVEN HYMAN: "Overprescription" may not be exactly right. My observation is that there are a lot of boys in suburbs who don't meet diagnostic criteria for ADHD but are on Ritalin.

DUMIT: Right, and that's what I wanted to ask about. If Ritalin does in fact help students do better in tests, and parents are having it prescribed for that purpose, maybe it's not the medication that's the problem here but this inappropriate "health care" use.

HYMAN: From its coverage in the media, people might expect that when you open a bottle of Ritalin you get a whiff of sulfur. It's just a medication; the issue is how it's being utilized and prescribed. There certainly are some very impaired kids who do much better when they're on Ritalin. And in clinical trials for well-diagnosed ADHD, the medication has been shown to be safe and effective. But as it's used more and more, people are developing a strong moral sense about it. They are acting as if the pill itself, as opposed to prescribing practices, is somehow defective.

What we know about ADHD:

- Bad outcomes: associated with academic and occupational underachievement, increased risk for substance abuse, arrest
- Associated with increased risk of other disorders (depression, anxiety, conduct disorder)

DUMIT: I'd add a third factor to that, namely the advertisements by the pharmaceutical companies, which pretty much directly target bad school performance as a symptom.

HYMAN: Absolutely. I would go out on a limb here and say that I think this kind of advertising is actually quite vile.

DUMIT: I have another question. When bad drugs—illegal drugs like amphetamines, cocaine, or ecstasy—are studied for effects on the brain, changes within the brain are seen as bad or neurotoxic. When good drugs like Prozac or Ritalin are studied for effects on the brain, changes are interpreted as proof of efficacy. So I wondered if we could nuance this. Do we see it as a neuroethical issue?

HYMAN: I would look at it very differently. I'd say that whether something is desirable or undesirable, or good or bad, can only be judged right now in terms of clinical outcomes, and that the observation of changes in gene expression or sprouting of new synaptic spines or synapses is really still not interpretable. We're trying to explain why certain drugs have very long-lived effects, but we can't tell whether something like new synapses is good or bad—with one exception.

When certain drugs lesion certain neurons—kill neurons—we have a good prima facie case that those long-term changes are not desirable. For example, the fact that ecstasy-like drugs are applied as off-the-shelf neurotoxins in the laboratory—you use it if you want to kill serotonin neurons—suggests that this drug is bad. We wouldn't need a big ethical conversation about that.

ANITA SILVERS (San Francisco State University): I have a question for Erik Parens, and it relates to the very good question Judy [Illes] asked this morning about whether or not we'll need to rethink the notion of "normal." You argue that we ought to be concerned about being complicit with unjust norms. I've encountered people who think that the treatment/enhancement distinction itself is complicit with norming species-typicality—after all, "treatment" is usually understood in terms of bringing individuals up to species-typicality, and "enhancement" as improving them beyond what's species-typical. Some people have argued that to use species-typicality as a norm is unjust. If that is so, isn't the distinction between treatment and enhancement itself complicit with an unjust norm?

ERIK PARENS: Anita, I have tried to learn from you that the treatment/enhancement distinction could inadvertently valorize the concept of normality, insofar as the argument seems to be that treatment ought to aim at normality. This is why I tried to underline that we use it only to begin the critique of some social practices and to begin to affirm the fact of variation.

So, yes, you're right. Any distinction can be put to very bad purposes and could, inadvertently, harbor premises that could be construed as problematic.

It seems to me, though, that the people wielding the treatment/enhancement distinction are also worried about what you're worried about. They're worried about not affirming a variety of ways of being.

Maybe we need to redefine what "normal" is because, for example, there are deaf people who want to have deaf kids.

Actually, I'd like to try to respond to what Judy said this morning, which was that maybe we need to redefine what "normal" is because, for example, there are deaf people who want to have deaf kids. I'd say, why not just affirm that there is a variety of ways of being, some of which, statistically speaking, are abnormal? Let's try to remember the difference between normal in the normative sense—the evaluative sense—and normal in the statistical sense. Being deaf, statistically speaking, is abnormal. Yet it's very important that we all learn how to say it is one of many ways of being and it is good, as good as any other way of being.

PAUL ROOT WOLPE: The argument about species-typical functioning in the treatment/enhancement distinction almost always comes up in discussions about what we should pay for in medicine. That is, we don't want to spend our health care dollar on cosmetic medicine, because it's not a treatment, but we do want to spend it to cure cancer. So where do we draw the line when we get to those things that could be considered treatment or not treatment? That's where the species-typical functioning argument happens.

It does not happen—or if it does, it's easily refuted—in the broader conversation about what is right and wrong. That is, the general social question of whether, for example, we should allow deaf people to have deaf children if they want to is a different issue from the question of what we should spend our health care dollar on. So we have to be very careful, when we ask the question about species-typical functioning, to note the realm we are talking about.

SILVERS: It's because we define deafness as a disease that we're in fact willing to spend our health care dollars on preventing deafness. I'm not arguing, by the way, that we oughtn't to spend money preventing limitations, but I am arguing that species-typicality is not a standard we ought to be using. We ought to be using a quite different functional standard, because people who are atypical in various ways may be as functional—or more functional—than species-typical people.

PARENS: Yes, I completely agree.

To even talk about a treatment/ enhancement distinction, it seems to me, is misleading because it leaves out multiple nuances and complexity.

MARY MAHOWALD (University of Chicago): Erik's discussion posed the treatment/enhancement distinction as a dichotomy, but there is actually a continuum in the demand for medical services—from severe disease states at one end to totally nonhealth-related uses at the other. And a whole array of conditions span that

Session III panel (l to r): Paul Root Wolpe, Erik Parens, Marilyn Albert, Steven Hyman, and Bernard Lo.

spectrum. There are also very different variables—economic, social, technological—that feed into them. So to even talk about a treatment/enhancement distinction, it seems to me, is misleading because it leaves out multiple nuances and complexity.

PARENS: Mary, let me try this again. All I tried to say was that the treatment/enhancement distinction is a tool that one might use for three purposes, which I listed. I said it could begin a conversation about what would go in and out of a basic package of care. It could begin a conversation about medicalization—about putting medicine to purposes that make us nervous—and it could begin a conversation about the kinds of natural variations that we want to affirm. To begin some conversations—that's all I think it's good for. It certainly can't be the end of the conversation. The fact of the matter is, you're absolutely right.

HYMAN: But maybe it's the wrong place to begin. For one thing, I hope we would agree that the boundaries between health and illness are partly socially constructed. For another, if there were public health people here, they'd be frothing at the mouth, because they would tell you that their goal is to have population-based prevention. And from that point of view, one person's enhancement is another person's prevention. For example, many people in this room are probably on a statin as a matter of prevention. If you're not, clinical trials show you have an increased risk of stroke and myocardial infarction, whatever your cholesterol is.

What is species-typical? It depends on the country you're in, the diet you're on, and so on. But even if you were foraging for roots and berries in some desert, your cholesterol would still be too high to prevent disease. All I'm saying is that the treatment/enhancement distinction has limits because . . .

PARENS: You used it yourself. I mean, your talk depended on it.

HYMAN: Yes, it did, in part, but . . .

PARENS: The whole of it!

HYMAN: . . . what's missing in that distinction is the issue of early intervention and prevention, which we can't call either treatment or enhancement, and that's where the whole world of public health resides.

From the Floor (Unidentified Speaker): I might be wrong about this, but I believe I've read that lower cholesterol is associated with higher incidences of suicide.

HYMAN: That has not stood up in more recent clinical trials, [laughter] though I kind of like it because I enjoy eating a lot of meat. There have subsequently been very much more substantial clinical trials, which show no such correlation. You can't be too thin or have too low a level of cholesterol.

FROM THE FLOOR (Unidentified Speaker): I'd also like to make the point that the presenters here are talking rather casually about altering human nature. My own opinion is that one of the best and most ethical things we can do, as scientists and especially as neuroscientists, is to take a very cautious approach and advise the public to be wary of alterations that involve any kind of invasion, whether physical—through such silly ideas as those implanted chips—or pharmacological.

I mean, it seems to me that we need some dimension of humility here. We ought to be realizing that we're the product of nearly 4 billion years of evolutionary history and that what we have in us is designed as a balance against variations of circumstance and reality—human beings are in fact a kind of all-purpose organism. Maybe we could make ourselves stronger in one way, but we'd lose something in another.

Arguments about evolution perfecting us are problematic because, for one thing, evolution didn't design us to live on today's diets.

HYMAN: Well, first of all I don't think we should be cavalier; whenever we are, we invariably pay a price. We would have to do careful clinical trials. But I'd also say that arguments about evolution perfecting us are problematic because, for one thing, evolution didn't design us to live on today's diets. And what's so perfect about many of our forebears not living past the age of 40 or 45? So I think it's a sort of naturalistic fallacy to bring evolution into this. I would prefer to say that we need to balance our views of a just and fair society against individual rights on the one hand, and weigh risks and benefits as best as we can determine them on the other.

You know, at one level, when I was a public health official I worried a lot about all these kids getting Ritalin. But on another level, we live in America, and some set of parents or some individual says, "I have a lot of distress. I don't meet your diagnostic criteria for depression. But Prozac makes me feel a lot better and helps me do a lot better in my life. How dare you take it away from me?" So I think we have to balance all these very difficult issues. We can't turn to evolution.

MARILYN S. ALBERT: I think it should also be said that Americans, and people around the world, are now voting with their feet—at least with respect to concerns like cognitive impairment. All kinds of things are being sold in health food stores, and advertised in newspapers and magazines, that nobody has any knowledge about. We see a lot of patients in clinical practice who are taking ginkgo or any of a number of these substances, and they have no idea what the side effects are. So people are desperate for alterations in their current status and desirous of enhancement. And I think our job is to give them guidance. Telling them to wait for the clinical trials, I can tell you, isn't going to work.

WILLIAM SAFIRE: Dr. Wolpe, in the first fourteen minutes of your presentation it seemed that the marriage of physiology and technology was taking off like a rocket and there was no stopping it, and we ought to get used to it. And then, at the last minute, you seemed to veer away from that and cast aspersions on it. So I just want to know, Did I fall asleep in the last minute? And what is your conclusion about the ethical considerations of this forthcoming wedding?

WOLPE: The problem is that I gave you only fifteen minutes of a half-hour talk, Bill. I was trying to say that what's going on now in the laboratories—such as people implanting chips in animals, and in themselves—is creating a new set of relationships between human beings and technology that may evoke new ethical questions, which we've only been dancing around the edges of here.

That is, by talking about specific technologies and talking about Ritalin or some other specific psychopharmaceutical, we're dealing in incremental ethics, which often leads you somewhere you don't want to be. At some point you have to step back, take a view from 30,000 feet, and say, "You know, all these things we've put together create a certain kind of fundamental shift, and we need to examine it."

I was also trying to say we have to somehow merge—or marry, if you will—both sides of human existence, which right now we tend to treat very differently. And this is not a matter for scientists alone. We sit here and talk about technological capabilities, but when we actually go down into the lived experiences of human beings, they often interpret those things, receive those things, place those things in cultural contexts very differently from the way that scientists, ethicists, and philosophers originally conceived. For example, deafness as a culture was not predicted by the scientists who were creating or exploring treatments for deafness. It was an emergent property of the deaf community.

Deafness as a culture was not predicted by the scientists who were creating or exploring treatments for deafness. It was an emergent property of the deaf community.

HYMAN: Paul, do you think that's just? Do you think it's okay for a deaf parent to want to raise a deaf child if the child has a chance not to be deaf?

BERNARD LO: And let me just spice things up a bit. How is that example different from a Jehovah's Witness parent who says, "Don't transfuse my child, because she'll be eternally damned"? Society steps in and says no. In this situation, we'll get a court order and transfuse. So where do we draw the line?

WOLPE: Jehovah's Witnesses will say they care about their children as much as anyone else, but the reason they don't want to get blood products is not because they want their kids to die but because they think that eternal damnation is worse than physical death. And deaf people will tell you they care deeply about their children, but deaf culture is a rich and unique culture in which they're deeply embedded, and they believe it's the best thing for their children. A society has to decide how it wants to balance individual liberty against some set of social standards about what is tolerable and intolerable behavior, and this is always a tough decision.

What do I, Paul Wolpe, sociologist at Penn, think about that? The answer is: Who cares? Each of us has only our one 300-millionth piece of that pie to contribute. And ultimately, it is not a decision that any particular ethicist or physician ever makes anyway; it's an emergent property of a social conversation that we all participate in.

From the Floor (unidentified speaker): I want to follow that up with a question I had originally intended for Dr. Albert. The intention in treating Alzheimer's disease is to restore decision-making capacity. But in research, and perhaps someday in clinical practice, we're liable to encounter patients who say, "No, thank you." And this would involve, I think, your issue of the balance between individual liberties and what we consider to be norms. So how much parentalism are we willing to assume there?

ALBERT: We haven't considered it so far because we don't yet have effective treatments. Overall, it's pretty clear that people are desperate to prevent a decline in decision making. But when the disease has progressed very far, they know that the likelihood of returning to their normal self is essentially nil. And to me, that's similar to people who have lived to be very old and are cognitively normal but who have undergone enormous physical decline.

You might offer them a treatment that would make them a little bit better, and they might actually refuse it.

Many of us know of such circumstances, and we generally believe that individuals who are in that position—and who have complete cognitive capacity—have the right to choose not to be a tiny bit better than they might be, because they feel as if even that state isn't good enough for them.

From the Floor (unidentified speaker): There's some precedent, though, in that society has developed a good position on the treatment of patients who are potentially suicidal. It's pretty clear that profoundly depressed people very often refuse treatment, yet we as a society make the decision that they should be treated. There are laws that say if you really feel the person's a significant threat to his or her own survival, you can force treatment. And once treated, they're completely different, and they'll often say, "You saved my life, and I feel a whole lot better about it." I think this is an example of what Patricia [Churchland] said earlier—that as the treatments become more effective and we understand the disease better, we'll be able to develop policies on which there is universal agreement.

It's pretty clear that profoundly depressed people very often refuse treatment, yet we as a society make the decision that they should be treated.

ALBERT: I agree, certainly with regard to Alzheimer's disease. What we need first are more effective treatments for it, and then we'll make the choice for life and quality of life. At the moment, unfortunately, we don't have those choices with respect to most neurologic diseases.

FROM THE FLOOR (unidentified speaker): I wanted to say something in response to that last comment [from the audience]. The fact is that some mental illnesses are terminal. It isn't true that everybody who is treated for suicidal feelings doesn't then—after maybe fourteen or eighteen tries—end up committing suicide. I think we have to recognize that our medical options can only go so far.

I also wanted to say that every time we draw a comparison between Ritalin and some educational enhancement like Stanley Kaplan or Germantown Friends, we're forgetting that the United States public education is seen as a public good that's free to everybody up to the age of 16. By contrast, we don't have a universal right to some basic, minimum level of health care, upon which Ritalin or some other medical enhancement could build. That's an extremely important distinction, I think, and the comparison fails there.

I have one other thing, just to go back—sorry, I've saved all this up. When you're giving the example "Should deaf parents have the ability to give birth to a deaf child?" are you talking about preimplantation diagnosis, or are you talking about not giving a child a cochlear implant? There's a huge difference.

WOLPE: And one of the differences is that the technology for cochlear implants, when you've talked to people who've had them, is not perfect—and it turns people into patients. Also, the technology's performance, such as it is, also depends on whether the implant is prelingual or postlingual; when teeny babies have the experience of hearing speech in the ambient environment, that's how they learn to talk. So parents have to choose which community their child will be cut off from—the hearing or the deaf community. There's some cause to think that the deaf community, given its culture and unity, is something to be missed.

The technology for cochlear implants, when you've talked to people who've had them, is not perfect—and it turns people into patients.

But there will be a deeper and more trenchant issue once we have a "cure"—once cochlear implant technology is nominally perfect—because right now it is exactly the poor functioning of people with cochlear implants that has allowed this sort of compromise to exist. Society, given the meager alternative it can offer, is allowing deaf parents (and hearing parents, too) to decline having their children get cochlear implants. Once we have a technology that functions well, however—once there's an organic cure rather than a technological fix—it will be a very interesting conversation we're going to have, as a culture, about whether we still permit parents to decline.

STEPHANIE J. BIRD (MIT): Whether or not cochlear implant technology is perfect, there's a larger question: Isn't it misguided for us to impose our own values in picking the community that people should belong to? That is, we in this group likely agree on the correct solutions, and the proper ways of imposing them. But those views are colored by our values.

So I think we really need, first of all, to recognize that we have those particular values and that they are embedded in our science. And then we need to develop procedures for looking at research findings and their potential applications not only for how we think it would make the world a better place, but for whom, and at what cost. We have that responsibility if we're really going to look at the ethical implications of the use of neuroscience seriously and not as elitists.

ALBERT: It seems to me that the only thing society has agreed to take a stand on is life or death. So that's why it's related to the people who are Seventh-day Adventists, that's why it's related to treatment of people who have severe depression and are at risk for suicide. But society hasn't decided to take a stand on whether or not you should treat children in school who could be helped if they took a particular medicine.

There is little appreciation of the complexity of the issues we're talking about. I agree—as a community we don't ask enough questions.

BIRD: Well, we certainly know, as Art Caplan noted earlier, that women often have amniocentesis because society (that is, the medical community, their family members, friends, and acquaintances) expects them to, even though it is not required by law. And, as someone else pointed out, there is little appreciation of the complexity of the issues we're talking about. I agree—as a community we don't ask enough questions.

HYMAN: In essence, this very impassioned discussion was all from the point of view of the parents. But the issue society is trying to balance, whether it does it well or not, and whether it has the right to or not, is whether the parents are the only parties who might speak for the child's subsequent desires. I don't think we can really answer that, but to me it remains a very important ethical question.

From the Floor (unidentified speaker): I thought I detected a convergence between two of the issues being discussed—one hypothetical, the other developing—on how we might apply the ethics of interventions. It came from both the race-discrimination pill and the cochlear implant. Interestingly, they both create what I think I can call a moral hazard problem. That is, in each case, the partial adoption—and even, in one case, the complete adoption—would lead people to abandon an effort that eventually could have been much more efficacious or broadly applied.

So if, for example, we accepted the idea of a race-discrimination pill, this might discourage society's more sustainable efforts to end discrimination altogether, as someone pointed out earlier. And I suspect that members of the deaf community, bound together by considerable effort to develop an alternative language that they share, might feel that the community's integrity would be damaged by even the partial adoption of successful cochlear implants. I just wanted to throw that out to all of you to see if it sparks anything.

HYMAN: The moral hazard problem is really very interesting and could probably take us off in directions that we ought not to go in. But I'd mention that in recent history there was a conviction in the psychiatric community that psychotherapy was superior to psychopharmacology. This was a deeply held set of beliefs that mental disorders came from unresolved conflicts, and that if you treated mental illness with medication, it was like treating pneumonia with aspirin; you would be papering over these internal conflicts and they'd rupture in some other place. There were even some famous cases of individuals being hospitalized to keep them out of the hands of psychopharmacologists, which ended in some pretty devastating lawsuits.

I think we're always suspicious of any shortcut in the relief of suffering, even if it makes us healthier.

I don't think we can generalize from the psychopharmacology example, but it turned out in reality that medication and psychotherapy can work synergistically or either/or. And the reason I'm raising this is that we still have an often-unexamined notion that something we work and suffer for, whether it's psychotherapy or exercise, is invariably going to be better than some easy solution. Perhaps it reflects our Puritan background, but I think we're always suspicious of any shortcut in the relief of suffering, even if it makes us healthier. So, for example, we still put people on diet and exercise regimens to lower their cholesterol even though we know it doesn't work very well and that in the long run people don't comply. But the individuals must suffer, and other approaches must be shown to have failed, before we use medical intervention.

PARENS: I'm the one who made the argument, kind of quickly, that in taking the discrimination pill there was a risk of becoming complicit with the discriminatory norm. One of the values there, of course, is that rather than change our bodies or anyone's body, we have to change our attitudes toward people whose melanin levels differ from that of the dominant group. But in the deaf case I think it's a little more complicated, because there are at least two values that we want to affirm.

The first is difference. The deaf culture is rich in many ways, and it makes a whole lot of sense to me that deaf parents would want to have kids who are deaf. Then there's this second value—among many—that parents have the liberty to bear the kind of child they want. But still, we have to think about the child's range of possibilities for the future. So it's very hard for me to put together those two examples that you interestingly juxtapose. To be honest with you, though, I'm not sure what the hell I think about the variety of deaf-issue cases, and I've been thinking about them for a while.

WOLPE: I'd like to borrow a phrase from Art [Caplan] that might help us. He talked about maximizing human flourishing. Is there anybody here who's against maximizing human flourishing? Okay, so we're all for it. In the deaf case, too, the issue is which of those two paths maximizes the child's human flourishing. And what makes the problem so difficult is that both sides are arguing for their own maximizing path, and there are two different visions of just what is being maximized. The only way you can win the cochlear implant argument is to discount the deaf-subculture argument, but as good liberal multiculturalists we don't want to do that.

PARENS: It would seem that as people we don't want to do that.

WOLPE: There are many places where they would want to—where that kind of subculture integrity is not something they grant to people. In this culture we do, and I believe it's the right thing.

PATRICIA S. CHURCHLAND: I don't really understand the argument against the race-discrimination pill. I don't understand at all why I would be complicit in racism were I to advocate taking such a pill. So let me change the example a little bit. We know that pedophilia is very difficult to deal with, and that offenders are usually repeat offenders. Now suppose neuroscience develops a pill that changes the brain in such a way that pedophiles no longer have that desire. Would you then say to me, "Well, you probably shouldn't advocate that, because it makes you complicit in pedophilia?"

Suppose neuroscience develops a pill that changes the brain in such a way that pedophiles no longer have that desire.

And the other thing is that when the number of people who are on Prozac comes up, it's often in a very touchy frame of mind, as though this is really a terrible thing and somehow we're not being true to our existential selves—that if we were to just agonize our way through our grim periods, then somehow we'd come out more in touch with our body or something. I don't really get that either. I know people who are severely depressed, but they think it would somehow show a failing of moral character to take Prozac. And that's partly because they hear—from ethicists, on television shows, wherever—that they're supposed to kind of keep their brain pure.

I'd suggest that all this is not just puritanical; there's something deeply "flimflamy" about it.

HYMAN: I know of only two instances in public health when interventions were actually stopped because they were said to be complicit in perverse values. One was needle exchange to prevent the transmission of HIV. This program was not permitted by the Clinton administration because it would presumably send a message that we were complicit with IV users of heroin and cocaine. More recently [Surgeon General] David Satcher got in a lot of trouble with the current administration after he wrote a report on human sexuality; it said that in sex education, besides teaching abstinence, we should also teach about condoms and other safe-sex behaviors. Again, this was interpreted by some people as being complicit in perverse values.

PARENS: Regarding the race-discrimination and pedophilia pills: I want to point out that it makes a difference how we respond to a problem like discrimination. As far as I understand, racist people can learn,

Problems:

- Community treatment is not as good as it should be (MTA (1999)
- A substantial number of children who receive stimulants do not meet criteria for ADHD (Angold et al. 2000): *This is a problem with healthcare, not with the medications.*
- A substantial number of children with ADHD are not treated.
- Schools are playing a controversial, often heavy handed role in treatment decisions.

they can respond to reason, they can come to see the badness in discrimination, whereas it seems to be difficult for pedophiles to see that pedophilia is bad. Now perhaps I don't understand pedophilia—perhaps I don't understand discrimination—but I wonder if we can distinguish between kinds of diseases. I think discrimination is one that we ought to respond to first with reason, whereas I'm resigned to the fact that with pedophilia we've kind of given up on reason and it makes sense to move to the mechanistic model.

ELLEN CLAYTON (Vanderbilt University): I'm going to make two very quick points. First, I wanted to point out a remark on Steve's slides that he skipped but that I think is very important: a major reason for methylphenidate use is that schools (illegally) insist on it. They send mothers to the clinic and say that if you don't put your child on methylphenidate, the kid can't go to school.

My second point is this: We have spent a lot of time talking about how the job of parents is to increase children's opportunities. Let me note that parents also frequently restrict children's opportunities. That's part of their job. And sometimes they even receive Constitutional protection for doing it, as in *Wisconsin* v. *Yoder*—Amish parents were permitted to take their children out of public school lest they be lost to the Amish faith. I think we need a much richer notion of what parents do, because a lot of it is to constrain their children's choices.

From the Floor (unidentified speaker): I just wanted to respond to Dr. Churchland's comments. If everybody ended up on Prozac and we were all happy and everything was great, we might not address some of the underlying causes of the rising rate of depression, which might have something to do with social inequity, which might have something to do with increasing stress as the workday gets longer, and so on. There are many, many reasons you might have an outcome like depression. Similarly with racism, which could be based on the country's power differentials, among other things. So by just treating the outcome, you might not get to any of the underlying causes, which are likely to be much more challenging and fundamental.

From the Floor (unidentified speaker): Also, how we evaluate performances or results depends on how they were achieved. So in my evaluation of an individual it makes a difference, I think, if the person overcame prejudice by taking a pill or overcame it through experience, argument, or thinking. It seems to me that there's a lot of room for nuance here, not in asking "Are we going to take these pills or not?" but in evaluating an enhanced performance—in our roles, say, as professor and judge—in light of the fact that it's partially generated by a pill or a device.

FROM THE FLOOR (unidentified speaker): I don't think it's a good idea to make a hypothetical example out of something as complex as racism. First of all, the notion that a pill could cure it is very hard for me to imagine. Racism grows out of a complex political/economic/ social culture and it serves very important dynamic purposes. I also find it hard to believe that if you could somehow take it away it wouldn't have other tremendous ripple effects in the society. I mean, poor classes have been manipulated by rich classes to use racism to avoid looking at other problems in the society, and so forth.

I don't think it's a good idea to make a hypothetical example out of something as complex as racism.

But the real issue in these sorts of cases, which nobody has touched here, is, Who's going to decide who takes this pill? And are people going to take it by choice? My guess is most people who have strong prejudices really believe in them and would resist tremendously somebody coming along and saying, "You have to take a pill so these classes of people, who you don't like, you'll now go out to dinner with regularly." Pedophilia is a different case, and I don't know enough about it to say how many of those who might be eligible for such a pill would voluntarily take it—

that is, would they indeed regard pedophilia as a curse, as something that they'd rather get rid of? Meanwhile, we certainly do know that a problem among people with severe mental illness is that they often don't take their medications, for reasons that aren't always clear.

LO: We are out of time. I want to thank our panel for a stimulating discussion, and also to thank them for adhering to the really nasty time limits I gave them.

Dinner Speech

Are There Things We'd Rather Not Know?

SUMMARY: Dr. Kennedy addressed three areas: "whether we humans are as neurophysiologically and behaviorally unique as we sometimes think we are"; "the customary and obligatory swing at free will," and whether our view of it will be altered as we learn more about the brain; and whether the government ought to regulate brain-related and other studies on ethical grounds. He noted that birds, bats, lions, and numerous other animals exhibit "human" behaviors such as altruism. "As we learn more and more about the neural and behavioral capacities of animals," he said, "the zone of what we think of as uniquely human is gradually shrinking." Meanwhile, given the complexity of the brain, free will is in no danger of being explained mechanistically, Dr. Kennedy said, though it's virtually certain that we'll learn enough to explain certain types of deviant behavior in neurological terms.

With regard to government regulation, he strongly believes that ethical decisions should be left to the researchers themselves, and that legislation like the Brownback Bill—which would criminalize certain nonreproductive stem cell experiments—though possibly well meaning, is highly inappropriate.

HOWARD FIELDS: My small role in this conference is to introduce the evening's entertainment. It's an honor for me to introduce Don Kennedy, and I'm going to admit something about our joint past. We both arrived at Stanford in 1960. He was an assistant professor in the biology department, and I was a wide-eyed medical student interested in behavior. By the time I graduated in 1965, he was already chair of the department. I've fallen further and further behind as the years have gone by.

About the time he became chair, Don shifted the focus of his intellectual efforts to areas in which biological science could directly inform social policy. And soon after that epiphany, he accepted appointment as commissioner of the Food and Drug Administration. He then returned to Stanford, where he became president in 1980. He weathered student protests, sometimes led by faculty; the devastating Loma Prieta earthquake; the Dingell Committee; and perhaps most frustrating, a consistent failure of the Stanford Hoops team to make the final four.

Despite these challenges, Don was able to revitalize undergraduate education, rebuild earthquake-damaged buildings, and win two College World Series—and I'm green with envy. Don is currently the editor in chief of *Science*, arguably the premier general scientific journal in the country. He has brought great intelligence, intellectual breadth, enthusiasm, and creativity to each of these major responsibilities.

Dr. Howard Fields, University of California, San Francisco.

But I'd like to give another introduction—a second introduction. This is to explain why I believe that Don's actual scientific interests make him an inspired choice to discuss some of the theoretical issues of this conference. So let's go back to the 1960s—to 1961. A year after Don and I arrived at Stanford, I took his course in neuroscience, which at that time was very much a nascent field dominated by electrophysiologists. The classic experiments of Hodgkin and Huxley, Katz, Eccles, Hartline, and Kuffler had been published and were making their way into our collective consciousness. They were elegant, they were rigorous, and they were seminal. And they described the basic properties of action potentials, synaptic transmission, sensory transduction; all the building blocks of the nervous system were beginning to appear in their outline form, and it was a very exciting time.

Now, Don came at the material from a different—and, I believe, more compelling—direction. His work began by defining a behavior, determining its biological significance, and only then did he proceed to reverse-engineer the circuits that generated the behavior. It's pretty well accepted that Don's a charismatic teacher, but my recollection is that this wasn't the reason I was drawn to work in his lab. I think it was because of his approach to the analysis of behavior based on neural function.

He used the reductionist method of single-neuron recording, which is still with us and is something that I'm still doing. But it was the organism-level focus that attracted me. The individual organism is the functional unit of animal evolution. Foraging, feeding, aggression, courtship, and migration are acts of individuals in a social framework, and they take place in an ecosystem. The construction of a biologically based neuroethics requires that we understand the value of the behavior to the species. That value is defined by the problems the organism has to solve—which, in turn, are defined by the ecosystem in which it evolved.

The approach of the biologist to ethical behavior takes evolutionary origins as its conceptual core, and we are fortunate indeed to have a biologist of Don's stature to address this subject for us.

DONALD KENNEDY: I want to thank Howard for establishing a context for what I'll discuss tonight. My own explorations have necessarily relied, as he told you, on a set of experiences as a neuroscientist and as a zoologist studying animal behavior, and some more recent ones as an administrator and publisher of research. I really have come to believe that evolution and our history are important shapers of what we are, how we think about ourselves, and how we think about what we know. And I hope that this context will intrude repeatedly on what I try to say.

Dr. Donald Kennedy, Editor-in-Chief, *Science*.

I'm going to touch on three areas, or "observation posts." In the first, which really is quite zoological, I'll struggle with the idea of whether we humans are as neurophysiologically and behaviorally unique as we sometimes think we are. Then I want to take the customary and obligatory wild swing at free will, asking whether we'll come to know so much about the brain and how it works that our notions about personal responsibility for one's own actions will be altered in some way. In other words, as we learn more and more, will we shrink the domain we know as free will so much that we stop seeing ourselves as responsible agents?

And finally, I want to enter—reluctantly, hesitantly—the policy domain and examine some areas in which government might seek to regulate studies of one kind or another on ethical grounds, as opposed to allowing scientists to choose what we as individuals care to examine.

So that brings us to the first observation post. What got me to study nervous systems in the beginning—and it's exactly what Howard told you—was an interest in behavior and evolution. Part of my work—the best of it, I must say, with Howard involved—was on the circuitry of fixed-action patterns in lower animals, where coordinated behaviors can be completed as a result of central pattern generators. These behaviors are often stimulated by a single neuron or a tiny cluster of them, and are completed without any help from proprioceptive or other kinds of sensory feedback. In other words, these are little motor scores. Another part of my work was an effort to understand how sensory systems extract useful information from the world, filtering stimuli as they do and equipping the behavioral apparatus to respond appropriately.

These central pattern generators, fixed-action patterns, and sensory filters that animals possess are the product of a long series of survival tests in which each line of contestants brought a different set of starting materials for natural selection to act upon. And much of what we want to discover about nervous systems and brains—learning, memory, sensory capacities, and other phenomena that at first seem inexplicable—can be understood in this way.

But what about the kinds of higher behavior that might interest a neuroethicist? Here I'll avoid some of those behaviors emphasized by sociobiologists (since they've become so politicized in one way or another) and focus instead on a key concept, called kin selection, that has been central to explanations of cooperation—the sentinel behavior of birds and meerkats, for example.

Natural selection ought to favor the propagation of individually disadvantageous behavioral genotypes only if they favor enough kin to offset the genetic losses to the individual. On that basis, the reproductive sacrifice of a worker bee is compensated for by the advantage conferred on her haploid sister, the queen; the infanticide by a newly arrived male lion in a pride is understood; and the warning cry of the sentinel meerkat is an attempt to improve the survival of his kin mates in the colony.

Well, there things stood. At the end of William Hamilton's remarkable work, and the extensions of it by Trivers and Wilson and others, it looked as though there was a suite of genetic tricks that could make us understand why animals that ought to behave selfishly appear to behave not-so-selfishly. Well, genetics, it turns out, doesn't get it all done. Some altruism is clearly present in animals, even in those that we regard as not only unlike us but downright unlikable. Take vampire bats, for example—a couple of people I talked to this evening were eager for me to cite vampire bats because they are so universally unloved, except by us. Vampires nest in colonial roosts, and they go out at night hunting for prey—a sleeping dog, or livestock, or, as horror films would have it, a beautiful woman.

Some altruism is clearly present in animals, even those that we regard as not only unlike us but downright unlikable.

It's quite obvious that this kind of predation doesn't always meet with success. I mean, you don't find a sleeping dog just anywhere. Sometimes bats score, and other times they don't score. A zoologist has now studied quite carefully the behavior of vampires that he individually banded to distinguish them as individuals within the colony, and he monitored them over long periods of time. It turns out that vampire bats, when they come home with a large blood meal, are apt to share it around: "There's more here than I can use, so please have some—and you have some, too." The researcher has also kept careful track of who the sharees are, how the sharers treat them, and how they then treat the sharers. And it turns out that—like humans playing iterated prisoner's dilemma games—bats reward individuals that have shared with them earlier, just as in tit for tat.

So there you are. We have to deal with different time scales, of course—phylogenetic, ontogenetic, real-time—but regarding real time we may well ask, Is the vampire bat exercising a form of moral decision making or just operating rationally in a survival game? As more and more is learned about the behavior of animals, it becomes—for me, at least—more and more difficult to get closure on a set of properties that are uniquely and especially human, and that can be unambiguously defined in that way.

The guy I did my Ph.D. with, Don Griffin, was a brilliant student of animal behavior. He worked out echolocation (sonar) by bats, for instance, and how migrating birds find their way home. But later in his life he became interested in what he called the problem of animal awareness. I think by that he meant consciousness—a view of how animals make decisions that was a little different from what his colleagues in Harvard's psychology department believed at the time.

That view no longer seems so outlandish, and in fact I'm coming more and more to agree with him. If you're skeptical on this point, I urge you to read a book by a remarkable contemporary ecologist, Bernd Heinrich, called *Mind of the Raven.* It's a truly extraordinary study that I think will make you come away wondering whether the only thing ravens don't do is talk. As we learn more and more about the neural and behavioral capacities of animals, the zone of what we think of as uniquely human is gradually shrinking. And as we learn more about how their brains work, it may well change our attitudes about how different we are from them, thus reducing our sense of being all that special. Maybe the main difference is that we have language, so we get to talk about our traits and to hold conferences on neuroethics.

As we learn more and more about the neural and behavioral capacities of animals, the zone of what we think of as uniquely human is gradually shrinking.

That takes me, I must tell you, into a space I'm not entirely comfortable with. I am no friend of the animal-rights movement, but on the other hand there's this awkward growth of knowledge that might in the long run change our view of our place in the living world. It may not change our view of how we deal with animals as experimentalists, but it will certainly change our view of the continuity of all living things—reminding us, perhaps, of Charles Darwin's remarkable insight on that Patagonian night: "We all may be one, we may be blended together."

So much for the first observation post. The second one takes me to the problem of free will, and I'm not going to do any better than—surely not as well as—many of you already did with that earlier today. But I want to make the point that free will might be the last line of defense, along with language, for defining what makes us uniquely human—at least, many people seem to think that.

Here we move onto more difficult ground. Suppose, for example, we were to learn enough about transmitter biochemistry and neuronal connectivity so that we could explain every single behavioral choice we make in those terms. Would we then be in a position to interpret departures from behavioral norms as deviant phenotypes? At Princeton, Jonathan Cohen is doing fMRI analyses of human subjects as they ponder moral choices. Suppose he eventually arrives at what you might call an architectonics of ethical decision making? Would that threaten our notion of free will?

Suppose that several decades of studies provide a much more complete picture of the way in which people make choices in ambiguous-stimulus situations. Would that further shrink the domain we think of as uniquely personal decision making? And what would we do with simple noninvasive-test results that revealed lesions responsible for unreliability or, more seriously, criminal inclinations? Would we extend the concept of the insanity defense to cover all phenotypes and abandon our contemporary concept of what constitutes free moral choice?

I raise these questions not because I have answers. The title of my presentation—"Are There Things We'd Rather Not Know?"—is itself an embarrassing display of my own ambivalence on this set of points. Two additional questions arise, which I ask in order. First, are we likely ever to get to that level of explanation? And second, if we did, how would we deal with the ethical sequelae?

I don't really need to deal with the second question, because I can't get past the first. I think the level of complexity in the brain is so great that I don't believe we will ever reach—certainly not soon enough to worry me—the depth of explanation that would constitute a serious challenge to the notion of free will or personal responsibility.

What we are quite likely to accomplish, however, is a level of knowledge that expands the range of neurological phenotypes that could be considered as exculpatory with respect to certain kinds of deviant behavior. That, I think, is surely a serious prospect—it is perhaps already here—and it will require some reengineering of the criminal-justice system. I believe, in fact, that one of the duties of people who are seriously interested in neuroethics is to predict what directions it might take and begin to prepare lawyers and others for the prospect.

One of the duties of people who are seriously interested in neuroethics is to predict what directions it might take and begin to prepare lawyers and others for the prospect.

At this point you might wonder why I haven't said much about psychopharmacology and the ethical issues that may flow from that kind of neuroscience. I suppose it's because, even though some future discoveries may create unforeseen challenges, I personally foresee most of the new discoveries as likely to provide net benefits. But there's a broader issue that's interesting to some—apparently including Francis Fukuyama—about whether our interventions into transmitter biochemistry aren't interfering with something called our nature. And I have some really strong views about this, because evolution has been taking an undeserved beating from Fukuyama, and even from one or two people at this conference.

To find a "natural" human, you'd have to strip away all of human culture, getting back to our distant ancestors with cranial capacities of about 500 cubic centimeters. You're back around 2 million years, because sometime between 2 million years ago and probably 800,000 to 900,000 thousand years ago, the first *Homo ergaster* or *Homo erectus* (depending on your preference) came out of Africa and began that migration across the Middle East. Even *they* had some tools already, which to me spells culture.

There's a very long history of interaction between the human brain and culture, which continues to this day. So it's a terrible misunderstanding to think that when human culture reached some particular level, natural selection just stopped. Natural selection went right on. In fact, what culture did was to give natural selection a kick start. It became explosive. And as a result you probably have the brain growing faster in allometric terms than any other organ in the history of vertebrate evolution.

So culture acted upon the brain; and the brain, of course, in growing in capacity and size, acted upon culture to create more of it. And you have explosive positive feedback, which has made us what we are. In other words, we've been tampering with our brains all the time. Now, there may very well be things we don't want to do to our brains, even things we shouldn't do to our brains. But to appeal to a natural state as justification for eschewing that kind of intervention seems to me to ignore some terribly basic biology.

For the third and final observation post, I want to allude briefly to a policy issue or two from my time in government and as a university administrator. In the latter capacity I found that discussions about scientific ethics sometimes devolved into concerns about end use. Independent of the researcher's intent, should some experimental project be interdicted if the results could, with some reasonably high plausibility, be turned to harmful purposes? I heard much more of this than I wanted as a university president. I had to umpire furious disagreements between those who believed that all information is good and those who believed that a possible abuse in the offing was adequate grounds for suppression. And I am mighty relieved to be out of it.

I had to umpire furious disagreements between those who believed that all information is good and those who believed that a possible abuse in the offing was adequate grounds for suppression.

I do want to note, though, that in general we tended to resolve those issues at Stanford—and I think they got resolved in much the same way at most other institutions—in favor of leaving the ethical decisions to the researchers themselves. We did this partly out of a conviction that research really is a form of speech, and that prior restraint is no less unattractive when it's applied to work in the lab than it is to the spoken word. Of course, there are research projects that are so plainly aimed at inflicting harm to others that they should be prohibited, but the bar for that ought to be set pretty high.

With respect to discoveries in the brain sciences, are there any legitimate concerns that could lead to methods of control that all of us in this room would find unacceptable? Some work, for example, might provide information that repressive societies could use for their own Orwellian purposes. Should we regulate such research, on the grounds that it might be malignantly applied *somewhere?* Or in our own society, would we condone research in which new methods are developed to make advertisers or others more able to manipulate subjects unawares?

We certainly don't need to see additional power added to the subliminal inputs we already receive from television. But in the main, I think the ethical decisions about what work ought to be done should belong to the experimenter, unless there is the serious prospect of external harm. If they're mandated by regulation, the doers never have to make their own ethical decisions; and I like circumstances in which people are forced to choose.

A real-world illustration is on its way, and I close with that. The Senate well may pass the Brownback Bill (S.R. 1899), which, like its earlier and already-passed House counterpart, bans efforts to clone human beings. That's okay so far, but wait. Suppose that someone in this room wanted to take a legal, pre-August 9, stem cell—that is, on the nether side of the Bush ethical boundary—take out its nucleus, and inject the nucleus from a cultured human brain cell to examine the effect of the reprogramming of that cytoplasm on subsequent differentiation. Suppose you wanted to do that. If you did, you could be sentenced under that law to a ten-year jail term. That's neuroethics, government style—scientists relieved of having to make their own decisions—and I think it's appalling.

Thank you very much.

FROM THE FLOOR (unidentified speaker): If you don't believe that animals have rights, has what you've learned about the way animals behave and think changed your view of how they ought to be treated?

KENNEDY: To a degree, yes. I didn't say that animals have no rights. I think they do have some rights—to humane treatment, for example—and those are important. The question is, Do they have the same set of rights that we allocate to human beings? And there I fall short. But I would certainly say that my experience in thinking about animals and their neural capacities and evolutionary history has made me respect them more. We are partners in experiments, for sure.

Session

IV

Brain Science and Public Discourse

Judy Illes, Session Chair
Senior Research Scholar, Stanford Center for Biomedical Ethics Department of Medicine and Department of Radiology, Stanford University

Colin Blakemore
Professor of Physiology and Director, Oxford Centre for Cognitive Neuroscience, University of Oxford

Ron Kotulak
Science Writer, Chicago Tribune

Michael S. Gazzaniga
Distinguished Professor and Director, Program in Cognitive Neuroscience, Dartmouth College, and Member, President's Council on Bioethics

JUDY ILLES: Yesterday we had an extremely productive day discussing important topics in neuroscience with respect to rationality, choice, social policy, pathology, and even the impending fine line between therapy and enhancement. There's no doubt from our discussions that these are intrinsically provocative topics. But we can also query their prevalence among our professional studies. So I'd like to quickly tell you about some pertinent data that my colleagues Matt Kirschen and John Gabrieli and I have generated.

We recently conducted a literature search of all neuroimaging papers published since about 1990—that is, since the genesis of functional MRI—that involve research using functional MRI either solely or in combination with the older neuroimaging modalities of EEG and PET. Our goal was to determine the volume of

Dr. Judy Illes, Stanford University.

studies over the past ten years, as well as any emerging trends in their neural/behavioral focus.

Our search returned 3,426 unique articles published across 498 different journals between the years 1991–2001 —an incredible explosion over time, both in terms of the number of journals and the number of articles. There was also a major shift over this period in the focus of these studies: the number of papers on proof of concept, methods development, and sensory motor abilities, for example, decreased. But there was a significant increase in the number of studies on complex behaviors and emotion—these includes studies of moral judgment, decision making, reward and punishment, self-monitoring, fear, lying, and deception.

In light of our discussions yesterday, you may or may not think that neuroimaging technology is able yet to give us meaningful information about complex cognitive behaviors in our daily lives. Regardless, the trends indicate that we are actually doing studies that go very much along that path. So we must pay attention to them.

We are still analyzing this immense database, but these trends do help inform the subject of this session: using and interpreting information about the brain and behavior, and sharing it with the public.

From the "Public Understanding of Science" to Scientists' Understanding of the Public

SUMMARY: Dr. Blakemore briefly described his long involvement in activities related to public understanding of science to show that he has "a foot in both camps." He then listed some of the reasons it pays to have a scientifically well-informed public, and he gave a short history of efforts in the United Kingdom to realize this goal. But he noted that these efforts had been frustrated by a series of science-based controversies, including mad cow disease and genetically modified foods. The result is that the public has become more (rather than less) skeptical about science, particularly concerning the perceived quality, or lack thereof, in the science advisory process. Recently, however, a paradigm shift has occurred that may ultimately raise the level of public trust and prove more rewarding for the science community as well. Whereas the process of the public's understanding of science was essentially one-way (from scientists to the public), the emphasis in Britain is now on dialogue and debate—two-way interaction between scientists and the public.

JUDY ILLES: Let me introduce our first speaker, Professor Colin Blakemore. In a profile published by *The Sci-*

entist this past April, he was described as an individual with boundless energy. He in fact has been flourishing in two parallel careers, one in neuroscience and the other in science communication. He has been described by the Royal Society as one of Britain's most influential communicators of science.

COLIN BLAKEMORE: Britain is ahead of the United States in very few areas of science, but in generating problems, controversies, and public confrontations involving science, we're really in the lead. I guess that's why I've been invited to talk in this session on the role of scientists in public communication of ethical problems.

Regarding my credentials in this area, I'm director of the Oxford Centre for Cognitive Neuroscience but am also chairman of the British Association for the Advancement of Science, the major national organization in Britain devoted to public communication. And as it happens, I'm also much involved with the Dana Alliance in Europe.

I did my first radio broadcast in 1976, when I gave a series of lectures, called the Reith Lectures, on BBC radio—six straight half-hour, no-illustrations presentations—which, amazingly, had a considerable audience. And since then I've been involved in what's now approaching 500 radio and TV programs, including a thirteen-part television series on the brain and mind. I say this not to impress you about me, just to let you know that I have a foot in both camps.

The agenda for the "public understanding of science"—that phrase is very much recognized in Britain—actually began about seventeen years ago. The two main issues that drove it were the recognition of the importance of having a scientifically informed public, given the speed of progress of science, and the need, in a democratic society, to involve the public in the decisions of politicians and commerce about how science should be applied.

Dr. Colin Blakemore, University of Oxford.

What are the advantages of a scientifically informed public? A very considerable one is giving people a better capacity to assess risks in their own lives. If risks are demonstrably so large that people should be protected from their own inclinations, then legislation tends to do that. But there is a wide range of public activity with an element of risk for which legislation isn't appropriate, and here is where people need to somehow be informed by scientists in order to make proper decisions about how to run their lives. And of course, some sources of risk come directly from technology and science.

In addition, it's important that people be able to assess the potential benefits of new developments in technology and therefore be better equipped to perform cost-benefit analysis in their heads; that way, they may make sensible decisions about what they want to do with technology.

Equally important is the empowerment of people to participate in public discussion and debate about where science should go and how technology should be applied. A broader, more metaphysical advantage is involving people who are not themselves specialists in science in the culture of science; this recognizes and endeavors to change the fact that while science makes a very important contribution to the culture of human society, only about 5 percent of the population is in any sense professionally involved.

And finally, from a political perspective, an educated public is more likely to be supportive of science-based policy, which, of course, is a basic principle for all developed countries.

In 1985 the Royal Society—the Academy of Sciences in Britain—commissioned a report, chaired by Sir Walter Bodmer, on the public's knowledge of science. It followed a paper in *Nature* that was a survey of the public's knowledge—or rather ignorance—of absolutely basic facts of science. The paper reported on answers to such *Who Wants to Be a Millionaire*–type questions as "Does the Earth go around the sun, or the sun around the Earth?" and "Do antibiotics kill viruses?" It turned out that people were abysmally bad at those things. I'm not sure it really matters very much that most of them don't know whether the Earth goes around the sun—it actually hinges rather little on their everyday lives—but it's an indication of the depth of the problem.

One of the Bodmer Report's major conclusions was a message to scientists: Learn to communicate with the public, be willing to do so, and consider it your duty to do so.

One of the Bodmer Report's major conclusions was a message to scientists: Learn to communicate with the public, be willing to do so, and consider it your duty to do so. The notion of scientists' *duty* was central.

In 1986 the Royal Society followed up on this report by establishing the Committee on the Public Understanding of Science (COPUS), at that time a very powerful organization. It was jointly administered by the three major scientific organizations that interface with the public and its leaders: the Royal Society itself, the British Association, and the Royal Institution.

During the following ten years the public-understanding-of-science agenda became deeply embedded in the scientific ethos of Britain. By 1995 another government-commissioned report, the Wolfendale Report, concluded that scientists and engineers in receipt of public funds have a duty—there's that word duty again—to explain their work to the general public. This report actually recommended that every holder of a publicly funded grant be required to participate in public activities. They would have to specify what public activities they had been involved in—whether it was local radio and newspapers, or national television, or whatever—before they would be eligible to apply for renewal. It was really quite draconian.

The research councils—the government funding agencies that disperse government funds—have all become involved in public-understanding-of-science activities. For example, the corporate plan of PPARC—the Particle Physics and Astronomy Research Council—states: "We believe that those engaged in publicly funded research have a duty to explain their work to the general public." On average, a research council spends about 0.25 percent of its total annual budget on such activities.

However, against that background of a decade of

increasing recognition that we have a duty to go out and explain our work to the public, a series of problems have confronted Britain and shaken to its roots the public's confidence in the scientific process. This public disillusionment applies particularly to the scientific advisory process regarding AIDS, mad cow disease (BSE) [bovine spongiform encephalopathy] and the associated variant form of CJD [Creutzfeldt-Jakob disease] in humans, embryo research, and animal rights (a topic in which Britain seems to specialize, unfortunately). GM [genetically modified] foods, of course, was immensely controversial, with that controversy even spreading back to the United States, where the technology had been much more happily accepted than in Europe.

Then too, there has been controversy over cloning and stem cell technology, cellular telephones and the new police version of telecommunications, the MMR vaccine (the triple vaccine for mumps, measles, and rubella, which, it's been claimed, might be associated with increased incidence of autism), and the foot-and-mouth disease epidemic in Britain just last year.

In this series of events, none of which can be directly attributed to the activity of scientists, a link was made by the media and by many members of the public that if things somehow go wrong with technology, it must be the generators of technology who are at fault. So the blame for many of these things was laid at the door of science—quite unreasonably, but that was a problem we faced.

Even more, these events shook the public's trust

in the scientific advisory process, especially for government. Why did government ministers continue until late 1995 to say that there was literally no possibility of a risk of transmission of mad cow disease to human beings? They shamefacedly had to admit only a few months later (in1996) that there was indeed very good evidence for such a shift. This series of events implied either that the scientists advising government were incompetent or that the process of transforming their advice into public statements was distorted. I'm absolutely sure that the latter is true, but the public got the impression that the former was the problem.

In the last few years, trust in scientists in general has—at least at times—been rather low. A poll by MORI [Market & Opinion Research International, a United Kingdom firm] in 1996 showed that 75 percent of the people had great or fair faith in scientists who were associated with environmental groups, while 45 percent had faith in scientists working in industry, and 32 percent had faith in government scientists. The public is more willing to trust scientists who don't, as it were, have vested interests—scientists associated with nongovernmental organizations that are committed to charitable acts and generally to fighting the establishment.

If you simply ask people to rank in descending order their trust of different professions... Close to the bottom are scientists.

Trust ratings consistently show that if you simply ask people to rank in descending order their trust of different professions, priests are always very high, doctors are high, teachers are high. Close to the bot-

Session IV presenters (l to r): Michael Gazzaniga, Ron Kotulak, Judy Illes, and Colin Blakemore.

tom of the scale are scientists. Then, I'm relieved to say, even lower than scientists are journalists, and right at the bottom—almost off the scale—are politicians.

Scientists' rankings in such surveys is very disappointing to those of us who've spent a good fraction of our lives trying to communicate with the public—in the hope that this would improve relations and understanding—and then discover that trust and confidence, if anything, has decreased during that decade. Interestingly, surveys around Europe show an inverse correlation between the level of public knowledge about science (at least with respect to those *Who Wants to Be a Millionaire*–type questions) and public trust in science. The more people know, apparently, the less they trust the scientists.

The more confident of those scientists who've been involved in the public-understanding-of-science process tend to say: "Well, this must be a transition period. It's a good thing; it's a sign of healthy understanding and inquiry. We'll move through it to the point where we can carry the public with us eventually." I hope so, but the results are nevertheless a reason for concern.

Actually, there has been a shift in attitude, post-BSE, with a 2000 report on science and society by the House of Lords' Science and Technology Committee. Headed by former Tory minister Patrick Jenkin, the study surveyed the successes and failures of the public understanding of science's agenda over the last ten years, and it came to the conclusion that despite all that effort, society's relationship with science is in a critical phase—there is a crisis of confidence in science among the public—but also that this crisis of trust has produced a new mood for dialogue.

In effect, the terminology has been turned around completely. "Public understanding of science" has become an even dirtier word, it seems, than its acronym (PUS), and now everybody talks instead about dialogue and debate—two-way processes of interaction between scientists and the public, rather than a one-way didactic presentation of the truth from scientists to ordinary people.

Why should scientists themselves be involved in this process of two-way communication? There are obvious advantages, the first being the authenticity of the evidence that people receive about scientific fact.

Involving scientists in presenting the results and evidence of science reveals the process of science.

Second, more subtly, is that involving scientists in presenting the results and evidence of science reveals the process of science. There's general agreement that members of the public know very little about how science really works. They tend to see the headlined breakthroughs and achievements. They expect science to come in irrefutable, uncontested truths, rather than in the turmoil of conflict and controversy that we know underlies the generation of, in the end, accepted ideas.

Showing scientists as people, showing they actually are human beings without two heads and without antennae sticking out of them, that they're the kinds of people who might live next door and have mortgages and kids and so on, is considered very important too. And so is policing the media.

The disadvantages are obvious as well: the opportunities given scientists to become showmen, to abuse the privilege of an audience to which statements can be made without peer review, to distort that evidence for whatever reason, political or professional.

And finally, there's the conflict of an involvement in public communication with the normal career path. It's generally recognized that activity in the public domain is still insufficiently incorporated into the career assessment of scientists, despite the fact that there's been a lot of lip service to its importance.

In any case, the emphasis in Britain now is on public engagement—not expecting the public to tell scientists what to do or expecting the public to judge the scientific facts and the worth of science, but encouraging the public to be involved in deciding where science should go, where its limits should be, and how it should be applied. The result, it is hoped, will be an increase in public confidence in the applications of science.

A variety of events in this spirit have been introduced over the last year or so by communications organizations in Britain. I'll just mention one of them: SciBars (or *cafés scientifiques*), which have spread very quickly around the country. These are free events in pubs and wine bars—anyone can walk in off the street, buy their own drinks, and participate in a public discussion about some scientific issue, usually led by a scientist. They've been fantastically popular and successful.

Finally, just to mention The Dana Foundation's involvement, it has very generously helped sponsor a new building associated with the Science Museum in London, three floors of which will be office space in which the European Dana Alliance will be accommodated along with the British Association. But another three floors are public space devoted to the science-public dialogue. Virtually any organization that claims it has science at its heart and wants to communicate—science and nongovernmental organizations, lobbying groups, and so on—will be able to have access to this space for genuine public dialogue.

Thank you.

Question and Answer

MARILYN ALBERT: Colin, aside from the issue of BSE and hoof-and-mouth disease, the list of crises that you mentioned is similar in the United States and Great Britain. Yet I think it's fair to say that there's less suspicion of scientists here. Maybe one of the reasons is that the involvement of scientists with advocacy organizations that have to do with disease—specifically with the translation of basic science to treatment—is particularly clear to the American public. Could you could say something about whether or not you think there's a difference between Great Britain and the United States in that aspect of public activity?

BLAKEMORE: I think that's quite right—it's one of the obvious differences between the U.S. and Britain. Another, and greater, cause of the difference in trust is that there is less of a tradition of openness in Britain and in Europe in general. We don't yet have a Freedom of Information Act, for example, though we're just about to introduce one. There's a far less open presentation by government of the process of government. Agency web sites and so on still include rather little of the information on which government bases its decisions.

This lack of openness and transparency is a major problem, though one that has been very much recognized now and is changing rapidly.

This lack of openness and transparency is a major problem, though one that has been very much recognized now and is changing rapidly. European leaders are coming to understand that when there are problems and members of the public can't get easy access to the reasons for the problems, they become paranoid and mistrustful of what they see are the responsible organizations.

MICHAEL WILLIAMS: Last year I was the chair of the AMA's Council on Scientific Affairs, and we faced these kinds of issues. But despite our openness, I think that large segments of the public remain skeptical. For example, the African-American community generally distrusts science, medicine, and researchers, and I think that clinical research as an enterprise has faced a number of challenges in the last several years as a result. We got burned very badly at Johns Hopkins last year in that arena.

My question for you is, Despite some differences between the American public and the British public, do you have any insights on things that are better or worse for us to consider in how to enhance this dialogue?

BLAKEMORE: I don't know to what extent the public in the States has had the opportunity to engage face-to-face with practicing scientists (except maybe during Brain Awareness Week, as far as brain research is concerned). This is a relatively new development in Britain which has been extremely successful. Every year we have a National Science Week, with hundreds of events and thousands of participants.

Meanwhile, the rate of science broadcasting on radio and TV has increased even further; it was high already. And the involvement of scientists themselves in broadcasting has increased tremendously.

These are a few areas, I think, where there have been real achievements. I don't know to what extent they're mirrored in this country, but certainly some lessons could be learned from them.

Let's Start with the Brain

SUMMARY: Mr. Kotulak noted how children's early life experiences, which a decade ago were deemed relatively unimportant in their education and subsequent fate, were shown by brain research to play a profoundly important role. The brain "uses experiences from the outside environment to form its circuits for thinking, memories, emotions, and other capacities," he said. "When stimulating learning experiences are sparse, the complex network of [synaptic] connections is sparse. The brain, it turns out, is a use-it-or-lose-it organ." Mr. Kotulak described the active interest of governors and other state officials, with the encouragement of august bodies like the National Academy of Sciences, applying brain research advances to childrens' benefit. And he offered suggestions for more and better interactions between scientists and the media in order to improve the timeliness and quality of information and ideas—regarding brain science and all other science—communicated to the public.

JUDY ILLES: It's my pleasure to introduce Ron Kotulak from the *Chicago Tribune.* He's the author of the book *Inside the Brain: Revolutionary Discoveries of How the Mind Works.* I'd like to quote from a paper he wrote on brain development in young children for a conference on research policy and practice: "Now we can see thoughts with new imaging devices that can spy on the living, working brain, and we can eavesdrop on individual brain cells to listen to their chatter."

Ron, I thank you for being here with us this morning at the Neuroethics Conference. How do we responsibly engage in discussion about those thoughts we can spy on? And what will the products of that discussion be?

RON KOTULAK: In 1990 there was no wave of public concern about the way young children were being educated, and most certainly there was very little concern about early life experiences—except Head Start, which as many of you know was withering on the vine for lack of support. Outside of a small group of dedicated people who pushed for better conditions for children, policymakers were mum and lawmakers kept the public purse strings tightly closed. There was no proof, they said, that early experiences could make a difference.

This attitude began to change dramatically as scientists provided evidence that good early stimulation builds better brains and that the lack of appropriate stimulation can actually harm the brain. As a result, many states, cities, and communities are now instituting measures to improve the quality of day care, expand preschool programs, offer help to new parents in raising their babies, and organize other efforts designed to improve the academic skills and mental health of children. These programs

Ron Kotulak, *Chicago Tribune*.

now exist in large part because of the role of the media in informing the public about the wonderful revolution going on in brain research.

Still, as evidence continues to mount about the brain's capacity to physically and chemically change in response to new learning, or the lack of it, society is facing growing ethical concerns. Have we done enough? *What* should we be doing? And how should we go about doing it? Such questions affect all segments of society: poor families whose children grow up in impoverished environments; middle-class families sending infants and children to second-rate day-care facilities; and rich families who trust their infants to the care of a nanny who lacks training in how to stimulate a child's brain.

How do the media go about covering this grand revolution and its impact on people? I got involved because of the growing interest in finding some kind of scientific answer to the question, Why do some children turn out bad? I was asked this question by the editor of the *Chicago Tribune*, who was appalled by the increasing rate of violence among young people.

How do the media go about covering this grand revolution and its impact on people?

The Tribune had undertaken a yearlong series documenting the lives and deaths of children age 14 and younger who were killed in the Chicago area in 1993. The findings were perplexing and frustrating. In case after case the lives of the children who met violent deaths followed the same disheartening pattern. Typically the children were born to teenage mothers. There was no father at home. The children were abused and emotionally and intellectually impoverished, and they lived in bad neighborhoods. They were failing in school. Some were both victims and perpetrators of violence.

Yet most children living in similar conditions did not turn out bad. What made the difference? Could brain research shed new light on why children behaved the way they did? The editor wanted to know, and to find an explanation I spent months interviewing more than 200 scientists from all over.

At first it seemed that the brain was still the black box it was always thought to be. Scientists could see what went in and what came out. Crucial knowledge about what was going on *inside* the brain, however, was still missing, though that was beginning to change with advances in genetic engineering, brain imaging, and molecular biology. Researchers were able, for the first time, to start to learn about how the brain works.

Critical to my research was the work of Peter Huttenlocher, an unassuming neurologist at the University of Chicago, who, almost unknown to anyone else, was counting synapses (the tiny connections between brain cells). They were so small and so numerous that they had previously defied a scientific census. Yet because they enabled brain cells to talk to each other and produce thoughts and memories, they were critically important.

Huttenlocher counted the number of connections between the brain cells of fetuses, newborn babies, children, adolescents, young adults, middle-aged people, and the elderly—these counts were all done, of course, through autopsies—and what he found was amazing.

The newborn's brain had more connections than the fetal brain. The baby's brain had more connections than the newborn's. The child's even more. The number of connections between brain cells continued to increase astronomically through childhood, adolescence, and young adulthood, when it then peaked, began to decline, and eventually plateaued, remaining at about the same level for the rest of a person's life.

This was an incredible finding, but what did it mean? Other scientists were beginning to find answers. Bill Greenough, a neurobiologist at the University of Illinois, showed that rats exposed to stimulating environments from birth had far more connections between brain cells than genetically identical animals raised in the usual seclusion of a laboratory cage. Importantly, the rats that had more connections were smarter.

The pieces began to fall into place. That explosion of connections after birth enabled the brain to learn from the environment. The animals that learned more also retained more connections. Those stuck in boring or depressing environments had fewer learning experiences and far fewer connections.

Other researchers found human counterparts. Children living in poor neighborhoods in Ypsilanti, Michigan, were divided into two groups. One received intensive intervention, which included enriched learning experiences and child-rearing training for the parents, while people in the other group continued to live in their usual way. Long-term follow-up studies showed that the children who got the souped-up enrichment had higher IQs than the control children, and they completed more schooling, got better jobs, had better marriages, and were less likely to be involved with the law.

It was a paradigm-changing concept. Most of the brain gets built after birth; it uses experiences from the outside environment to form its circuits for thinking, memories, emotions, and other capacities. When stimulating learning experiences are sparse, the complex network of connections is sparse. The brain, it turns out, is a use-it-or-lose-it organ.

These kinds of findings, repeated many times since and covered widely in the media, were the driving force behind early-education programs that took root in the mid-nineties.

What was trickier for reporters, though, was trying to find out what researchers were learning about the chemistry of violence—a political and social minefield. Because it involves studying the biology of behavior, many people were fearful that this kind of research could be used to discriminate against some groups, and perhaps be used for mind control as well.

In 1992, Fred Goodwin, director of the then prestigious Alcohol, Drug Abuse, and Mental Health Administration, created a firestorm of protest when he said that aggression in some people seemed to be similar to aggression in primates. He was referring to the similarity of certain chemicals in the brains of apes and humans. But many mistook his comment to mean he was calling some aggressive people apes. The heat got so bad that Fred had to resign his position.

I was investigating this kind of research but had reached a dead end. No scientist working in the field wanted to talk about it because of the obvious implications it might have for them. Key to breaking through this barrier that scientists had erected was the late, brilliant Markku Linnoila of the National Institutes of Health. He had new evidence linking the levels of a critical brain transmitter, serotonin, to aggression as well as depression. But he refused to talk. I needed somehow to break down the barrier and convince him that I would do a fair, accurate, and responsible story about his work and that of his colleagues.

One day I learned that Linnoila had given a speech on his research at a meeting here in San Francisco. I got hold of the people who recorded its sessions and ordered a copy of the one he spoke at. The talk contained good information about his research, but I needed more in order to make a whole story. So I transcribed his speech, sent him a copy of the transcript, and then called him and asked if we could now talk. He agreed. With that interview in my pocket, I was able to persuade other researchers to talk too. The resulting series of stories I wrote, called "The Roots of Violence," was as balanced as I could make it, so much so that I observed no hysterical reactions from researchers; in fact, many of them said it opened their field to the public in a way that took much of the emotional tinder out of it.

These early findings led to the development of Prozac and other drugs to treat depression by raising serotonin levels. The drugs have become widely popular, but they have also created new ethical questions. Many people are now worried that they may be over-prescribed, especially for children. Similarly, while other drugs may come along that help make people happy, will these agents also be used to make populations more complacent, docile, and controllable?

While the potential for abuse must be monitored, a bigger problem at present is that people are making insufficient use of these drugs, and other types of new knowledge about the brain, for their very positive effects. When leaders and policymakers are given the chance to be informed, however, they usually respond; it's easy to see that these advances can make a difference in how children learn.

The Education Commission of the States, made up of the nation's governors and state legislators who develop education policies, sent letters to all the governors, offering to send experts to talk to their key people. They had expected maybe a handful of governors to accept, and were surprised when almost all of them did.

Indiana governor Frank O'Bannon said, "This could well be the greatest challenge our state faces during my administration—to incorporate what we know about the importance of our children's earliest years into our public policy, to make preschool more educationally enriching, to make parents more aware of how they can make children smarter, to encourage innovative community-service projects focused on early childhood, to make it possible for parents to spend more time with their children, and to make it count."

Governor Paul Patton of Kentucky, an engineer, said, "I thought that we were born with a computer brain and the challenge was to fill the computer up. Now

I have learned that we're born with just a lot of the parts and the computer is built after the child is born."

Things seemed to be looking up for children, but in late 2000, Americans were doused with a bucket of cold water: they were told that they weren't doing anywhere near enough, and that neither were their institutions. The National Academy of Sciences, in a report titled "The Science of Early Childhood Development," scolded the nation for not helping its children and families prepare for the changing demands of life in the twenty-first century.

The National Academy of Sciences... scolded the nation for not helping its children and families prepare for the changing demands of life in the twenty-first century.

Zero to Three weighed in with another report, "What Grownups Understand About Child Development," which documented how little parents know about the learning capacities of their infants. A third report was released, by the U.S. Department of Health and Human Services. It was called "A Good Beginning," and it said that children's unpreparedness for school was increasing. It blamed the problem on a lack of emotional development and social skills.

This kind of interest and concern among governors and various august bodies may signal the beginning of a groundswell for programs to enhance the mental development of infants and preschoolers—like the one that occurred in earlier times, when America decided it was a good thing for children to have universal public education.

The media's job in the face of these problems is clear. It's not just to inform people about all the high-tech advances but to tell them that neuroscience research is also providing some simple and practical solutions. What works are things like stimulating emotional development through love and hugs, and enhancing intellectual skills through talking and reading to infants, right from the start. These things can be done in the here and now, and they work.

But while the media help, they can also harm: an unintended assault on the brain comes from television. Most people don't believe in censorship, yet we are faced with increasing amounts of violence in TV programming. More than 250 studies have indicated that watching too much TV, especially violent shows, influences many children to become more aggressive, though not every viewer is affected. Overall, violent TV shows may account for about 10 percent of aggressiveness in children. It's not overwhelming, but it is significant. Particularly worrisome is the latest long-term study showing that TV may be able to increase aggression in older people as well.

Meanwhile, the media's role in highlighting the contributions and positive applications of science depends heavily on the cooperation of scientists. Let me give you some suggestions that I've culled from my experiences and those of other people, like the *Los Angeles Times's* Lee Hotz here [attending this conference], for improving the lines of communication between scientists and reporters:

Scientists who are reporting new findings should be available for interviews. They should be available even if just on a background basis—to help reporters evaluate new material, for example. This is especially important when statistics can be used in misleading ways, such as in not distinguishing relative risk from actual risk. Scientists should try to speak in layman's terms, using analogies and metaphors when appropriate to make concepts easier to understand.

Scientists should try to speak in layman's terms, using analogies and metaphors when appropriate to make concepts easier to understand.

And they should think in terms of explanatory illustrations, such as graphs or drawings, that can help make complex ideas and results more understandable to the average person. At the *Chicago Tribune* we now have a huge department of artists who do nothing but try to make material, whether it's scientific, medical, geographic, political, or anything else, more accessible. The efforts of this pioneering group have been very successful.

I would also like to emphasize a very, very important point that Colin [Blakemore] made earlier. Too often the public greets a new scientific announcement as a final fact, only to become confused and frustrated later on when another report contradicts it. We've all heard stories about, say, a report in the *Journal of the American Medical Association* that suggests caffeine is good for you which is followed up the next week by an article in the *New England Journal of Medicine* that says caffeine is bad for you. And it goes back and forth.

This is very confusing to the public, and I think that's part of the problem we're facing. We don't have enough people putting this kind of broad and overwhelming information into focus, or emphasizing that science is not carved in stone but is a constantly changing process of discovery.

When I first started covering science—and I've been doing this for more than thirty years—I could write a daily story without much complication because I merely had to address what little was available. Today we're inundated with information coming from all over, and it's part of our job to try to put it in perspective and make some sense out of it. Meanwhile, I think it's part of the scientist's job to help us do that.

Thank you.

Question and Answer

SARAH CADDICK (Steven and Michele Kirsch Foundation): While I agree with you that scientists should be more cooperative with the media, one of the things that's missing in the media is reporting on the actual process of science, as Dr. Blakemore pointed out. This is not something that sells newspapers, but I think it would help in the public's understanding. I'd love to hear your thoughts on this.

KOTULAK: Actually, I think that the science process does sell newspapers. If you look at *Newsweek* and *Time*, their cover stories on science or medicine are their biggest sellers. And the well-done and popular Science Times section helps sell the *New York Times.* People are fascinated by it and they want to read it.

If you look at *Newsweek* and *Time*, their cover stories on science and medicine are their biggest sellers.

Sure, people want to know about cancer—the immediate things that are important in their daily lives—but they have broader interests. We have a brain that allows us to say, Where did you come from? What are you doing here? Where are you going? All of science is an exploration of these issues. And I think that people, whether they acknowledge it or not, are intrinsically attracted to that exploration.

Even better, editors have become much more aware of such reader interests. When I first started writing about science and medicine at the *Chicago Tribune,* I was pretty much alone on the beat and my stories had no guaranteed space: just about any of them could get kicked out of the paper if a fire broke out or somebody was murdered.

Today we have a staff of some eight people covering science and medicine, and our stories get a lot more respect and priority. If you talk to Lee [Hotz], you'll learn that the *Los Angeles Times* has about twelve people on that beat. So there is now an incredible public interest in what's going on in science, and it's more important than ever that we take a broader look and make sure that things are being put more into perspective.

For example, when you folks were talking yesterday about free will and numerous other issues, these were fascinating sessions because they were free-for-alls. It was fun to see the openness and the sometimes-contradictory discussions on such a wide range of contradictory ideas. It reminded me a little bit of the Asilomar Conference in that scientists were saying, Let's watch out for the dangers here, but let's also look at the promise.

The same kinds of breadth, tolerance, and pragmatism, I think, apply to the public. So I tend to be an optimist. I think that readers are almost as interested in the process as they are in the breakthroughs.

The Pope, the Rabbi, the Scientist, and the Neuroethicist: Who Should You Believe and Why?

SUMMARY: Dr. Gazzaniga offered some lessons learned from his service on the President's Council on Bioethics—a group of accomplished and likable but highly diverse individuals who have altogether different ways of seeing the world. He described the need to learn some new vocabularies and how to "think in public." Neither one is easy, but the greatest challenge he seems to have faced on the panel was "moral equivalency" with regard to cloning. People with strong beliefs tend to be tenacious in those beliefs—one person may see a blastocyst as a clump of cells, and another may see it as morally equivalent to a Henry Kissinger. Analogies, even facts, can help; but they only go so far. Under the circumstances, the best thing a person can do is proceed as tolerantly and flexibly as possible. "When you get into that room," Dr. Gazzaniga said, "you have to try to develop a rapport and see where the discussion goes." But scientist-participants might find the process a little easier if nonscientists were more aware of three things: scientists' passion for learning nature's secrets, their intolerance for sloppy work or sloppy thinking, and their basic skepticism—even, or especially, regarding colleagues' ideas.

JUDY ILLES: Our final speaker for this session, Dr. Michael S. Gazzaniga, is the author of several books, including *The Cognitive Neurosciences,* which has played a major part in defining that field, and *The Mind's Past,* which "discusses both the brain's illusion as well as its construction of personal identity and memory, offering clues to the puzzle of consciousness."

Professor Gazzaniga, the title of your talk suggests that the pope, the rabbi, the scientist, and the neuroethicist might all have very different views of consciousness and what makes us human. So you'll have to tell us: What is the role of each, and how do we ensure that each has a voice? Who should be believed? And why?

MICHAEL S. GAZZANIGA: The people of that title all have beliefs—*strong* beliefs. They have highly developed points of view. And normally they talk to those who believe them, which makes for an anxiety-free day.

But when you put these people or their representatives around a table, you have a different game. My mission here in the next fifteen minutes is to give you a case history of such a process—something that happened to me—and to tell those neuroscientists and budding neuroethicists among you what's coming if you're called upon to take part in one of these discussions.

Dr. Michael Gazzaniga, Dartmouth College.

So I'll touch on my experience to date being on the president's panel [the President's Council on Bioethics]— not so much regarding the topic, but its process and interactions.

When you get a call from the

White House, no matter who you are and what your politics might be and what your personal views are, you take it. And when you're asked if you'd be interested in doing the assignment, aside from the fact that Hanover winters are cold and it looks like a good way to break them up, you of course accept it—and with a sense of pride and duty. If your country's leaders want you to help think them through something, you go do it.

The first thing you discover is that you don't actually know anybody on this committee—save for one person, in my case—so you have the obligation of getting to know them and learning how their minds work.

There were dire predictions in this regard. Our good friend Art Caplan said that this was a council of clones about cloning. But he was a little off on that. I've come to know these eighteen people and they're all very good, very smart, and they believe what they believe with great intensity. And I respect that. I also have a personal test for people: Would they pull you out of a foxhole? If so, then I count them as friends. All members of that group pass the test.

Another thing you have to deal with, almost immediately, is adapting to thinking in public. Now, most of us go to seminars in our fields, and when someone is veering off into an idea that you think is nonsense you say so—usually in very frank terms, like "Get a life!" or "Where have you been?" or "That's medieval!" Such things come to mind during these panel sessions, but because you're in the public eye you have to learn to speak with a different sort of veneer. And in the long run that's good, because it keeps discussion at a civil level, which is to the benefit of all.

You also have to learn all these different vocabularies. You're a knowledgeable scientist, but it turns out you don't know much. You know about what you do, but the next table over is full of information that you don't know how to use. If you ever do need to use it, you have to go study it.

Adding to that is the language of the ethicist and moralist—something you really have had no experience with if you've been a laboratory scientist all your life—and I recommend going very slowly in that area. I can never keep straight whether it was Hume or Kant or Hobbs or Aristotle who said X, Y, or Z. Maybe I should make a little cue card so that I could look down and remember which one to quote when I'm trying to sound like I'm historically in tune.

He discussed a religious cult in Minnesota that predicted the world was going to come to an end at twelve o'clock one fine night. Well, what happened when it *didn't*?

When you're in this committee's meetings you see not only that beliefs are strong but that it's very hard to rebut them, even with solid facts; such beliefs are in all of us, we've reflexively developed them, and they're seemingly there to stay.

Leon Festinger, the great psychologist, studied these kinds of beliefs years ago in his book *When Prophecy Fails.* In particular, he discussed a religious cult in Minnesota that predicted the world was going to come to an end at twelve o'clock one fine night. Well, what happened when it *didn't?* Did the people say, "Aw, this is a crazy religion, we're outta here?" No. They adapted their beliefs, citing some error of calculation, and they readjusted the date. In other words, completely confronted with contrary information, they didn't change their belief system at all.

Scientists are not immune to this phenomenon. Everybody thinks when scientists are shown a body of data that disconfirms one of their beliefs, they simply say, "Oh, that's it" and then walk out of the room readjusted to the truth, and for life. But nothing could be *further* from the truth.

Kepler plotted the first three points of planetary motion and looked down at his desk, saw an ellipse, and thought it couldn't be right—the calculations must be wrong. He was a religious believer, after all, and God would have made the paths of these planets perfect circles. So he had to plot hundreds and hundreds of points before he finally, literally, had an ellipse drawn on his desk and he gave in to the data.

Cognitive scientists have a whole series of studies in which they take, say, preachers and scientists and put them in a conflict situation—where a particular piece of data conflicts with what they believe. It turns out that preachers are quicker to adjust their beliefs to new data than are scientists. We all have friends like that.

It turns out that preachers are quicker to adjust their beliefs to new data than are scientists.

I'll just give you an example of how quickly we develop this belief thing and how tenaciously we hold on to it. Studies have been done on people who buy lottery tickets that bear nonsentimental, computer-generated numbers. A guy has just bought a lotto ticket for a buck, and as he's leaving the store the psychologist walks up and says, "I'll give you two bucks for that ticket. That's a 100 percent profit in five seconds." The studies show that in fact the guy won't sell it for two bucks, four bucks, or five bucks. It has to get up to twenty bucks before he finally accepts the deal. This is preposterous, and yet we do it—there's a vast literature on the psychology of commitment and how it sticks.

So when you're sitting around a table with this new group of people, all of whom represent different aspects of the culture—theologists, humanists, pragmatists, scientists, physicians, and sometimes ordinary people who are just worried about the future—they are all deeply committed to their own particular ways of thinking. You're best advised not to try to convince them otherwise but to instead stick to what you know and try to think about the problem in a way that can maybe help the discussion along.

When the committee considered cloning, the big issue was moral equivalence. Some people were concerned that the blastocyst, which resembles a person not at all, could well have the moral equivalence of a Henry Kissinger. You point out that it is just a tiny clump of cells, as opposed to a full-blown, developed human being. That "perceptual representation" of a scientific point has impact.

You also offer other schemas for helping to think about the problem—for example, the transplantation model. We already have in our culture the acceptance of "brain death" as a standard, and we harvest organs routinely. Some 10,000 to 20,000 life-saving surgeries go on each year, and they are accepted by religious groups and other social groups. So, you say, what about this clone blastocyst?

Can't the people who are involved in generating the somatic cell and the egg say, We're glad to give this tissue over to science, as with whole-organ transplants?

But then you get into what's called the slippery-slope argument. People are worried that if we do this blastocyst thing, we'll slip right down into a terrible situation—a dystopic future. By that argument, when we lower the voting age from 21 to 18, we should be concerned about soon seeing 2-year-olds vote. You know, we're good at drawing lines, putting friction on that slope. That's what our society does, and whole professions are given over to it.

You can also point out that the vast majority of people in the world are actually for, not against, biomedical cloning. But these kinds of arguments, analogies, and even facts do not necessarily prevail in a group of non-like-minded individuals, even when they are all very smart and sophisticated.

Part—though not all—of the problem is that nonscientists usually do not understand the world of science and its practitioners. We need to improve our representation to the public, which thinks of us as nerds who do these bizarre Frankensteinian things. I think that failures often come when members of the public don't know three things:

First, they don't appreciate the moving experience of the scientist's actually discovering something and learning a secret of nature. That's a feeling all the scientists in this room know. They respect it deeply, they understand it, and it energizes their lives.

Second, the public doesn't realize that scientists

are ruthlessly conservative. They are people who absolutely cannot abide sloppiness in an experiment or a report.

Third, the public needs to see the difference between science and scientists. As we all know, scientists can be too excited about their work to put a perspective on it. So nonscientists are generally unaware that when one scientist is talking to the other, basically the listener is thinking, How is he wrong? What is the flaw in whatever it is he's saying? It's not that scientists are out there trying to push the baby over the cliff; they are just people who keep a brake on runaway crazy ideas in our culture. To communicate that notion is vastly important.

The public needs to see the difference between science and scientists.

What causes a group to fail to incorporate objective fact? Well, I think it's natural. You have to learn that it's normal for people to have strong beliefs. These beliefs are part of their whole personality and their life history, and when you get into that room you have to try to develop a rapport and see where the discussion goes.

What can be done to change those beliefs? It's the old need for time and education. I don't mean that the prescription for science to have a greater influence on ethical issues is just to report more science. We're already inundated with science reporting. I think that teaching and trying to educate the public more about what science is, what it does, and how it actually does its business will make people feel vastly more comfortable when the scientist talks.

Thank you.

Question and Answer

JOE DUMIT (MIT): We've heard a lot about the image of science, but the scientists getting a lot of press today are often working for corporations. So the students I teach often see news about science coming from someone in a PR role, as opposed to someone in a research role. Could you comment on how this changing notion of the public scientist is affecting the goal of better communicating the results of science to the public?

KOTULAK: That's a really good question, and in fact we've tussled with it quite a bit. Many of you can remember—I certainly do—that biology used to be a bench-top science and that biologists used to feel they were working for the great glory of discovering something new, which to me is the greatest incentive of all.

The federal government has helped move this trend further along by passing laws allowing scientists to patent the results of their federally funded research.

But along came the tools of genetics, and pretty soon biologists discovered that if they found a new gene or a new protein, it might be usable in the marketplace. There's a famous story at the University of Chicago, where they developed something—I forget the exact name—that worked in the marrow to stimulate blood growth. After the researchers published it, somebody else went and patented it, and now it's a multimillion-dollar product. As a result, most scientists now think really carefully about their work's commercial prospects. And the federal government has helped move this trend further along by passing laws allowing scientists to patent the results of their federally funded research in order to bring products to market.

The thinking is that all this helps the economy. I still have a problem with it, though, and I'm sure everybody else here has a problem with it. Where is that high-level purity that we're all thought to have? Still, I'm coming to terms with it. I realize that this is part of what's happening now, that we have moved—probably forever—away from that notion we had, if it was ever true, of pure scientific work. But now the approach is clearly: By golly, let's make use of it, and do it fast.

I have a colleague at work who still has not gotten used to this, and he rails about it almost every day. He says that the results of research done with public money should be public information, without their being appropriated for private profit alone. But for myself, I bend a little bit, because it's probably encouraging more people to get out there and find things. We're still in a transition period, though, and I don't know where it's all going to go.

DICK TSIEN (Stanford University): I have a couple of questions, and the first one is for Colin Blakemore. You've pointed out the importance of dialogue, and just to be provocative, I'm going to ask you whether you think it's possible that a better way of engaging in dialogue and engaging the public—besides scientists talking to the public and vice versa—is to have scientists conduct more debates with each other in public. It often works well, and I think the great success of your brief lectures might have been even greater if you had debated someone with a responsible but different point of view.

My question for Mike Gazzaniga—also in an effort to arouse an argument—is this: You may or may not have heard Jonathan Moreno argue yesterday that this whole emerging field of neuroethics is a distinctly American phenomenon. As a practicing scientist, I find that notion really weird and hard to accept. The beauty of science is that it's not British or American or German or Southeast Asian; it's international. The issues that unite us are far greater than those that divide us, and I would hate to see us start on a path toward science, or ethical aspects of science, being in any way branded chauvinistically as American. But maybe Jonathan was making a point that I didn't completely understand.

The beauty of science is that it's not British or American or German or Southeast Asian; it's international.

BLAKEMORE: I think your idea of debates between scientists in public, or between scientists and those with a different interpretation of scientific evidence, is a very good one.

We have to recognize that there are various agendas—subtexts—involved in this move toward dialogue. One is to genuinely give the public a sense of ownership of science. Now we could say, Why bother? It's supporters in industry, or it's politicians, or it's scientists at the universities who own the results of scientific efforts. But that's not true. Wherever your money comes from, whether directly from public funds or charitable foundations, or even from industry, in the end it has come from the public purse, from people spending money in order for scientists to do their work. So it is, in a sense, owned by the public.

Another subtext is to genuinely tap the common sense of ordinary people in arriving at ethical decisions about how science should be applied. It's all too easy to become detached, either by your own inflated view of the importance of your work or your commercial interest in it, from how ordinary people are going to react to that work.

The third, an important subtext that I mentioned before and so did Mike, is to increase public knowledge not just of scientific facts—the Trivial Pursuit–type facts of science—but how science works. That way, the public will not be frustrated when it sees different headlines in the newspapers on successive days about the benefits or adverse effects of caffeine, and people will not be confused when they see figures expressed in terms of probabilities rather than certainties.

So there are various agendas, some of which I think can be best informed by exactly the sort of debate you propose, others by just genuinely involving the public and discussing with scientists how they work and how their work should be applied.

ILLES: Mike Gazzaniga, would you like to respond? And then I'd like to ask Jonathan Moreno to respond as well.

GAZZANIGA: The fact of the matter is that the United States is the biomedical engine of the world, whether we like it or not. This comes up repeatedly in the cloning issue, where people say, If this is outlawed, it will be sucked up by Singapore, China, Japan, or England, and the job will get done anyway by others. That's just simply not true. Ninety percent of the biomedical enterprise is right here. Therefore we are confronted earlier with questions, and we also discuss them more freely than other countries do. That's going on not only in the cloning issue, but I think will occur in this new topic of neuroethics as well.

The fact of the matter is that the United States is the biomedical engine of the world, whether we like it or not.

But even though Americans are going to be the first ones out of the gate, by their very nature these issues will become international topics of great interest. And I wouldn't want to see a big committee managing them for the entire world; I can't imagine anything more paralyzing than that. So maybe we are the first ones out, maybe it is peculiar to us now, but that will change.

JONATHAN MORENO: Just a point of clarification. The argument wasn't that science is peculiarly American but that the way the discourse in bioethics is developing has been dominated by some values that Americans find especially attractive—particularly the notion that consent trumps other considerations. That was the view.

And I think anybody who's attended ethics meetings in other countries has noted that the discourse is quite different, and the subject matter and the points of concern are quite different, than in the United States.

MARY ELLEN MICHEL (NIH): Since we have people from the press here, I wonder if they could give us some insight about skepticism and the delivery of negative information. So often in the press, it seems to me, every discovery is interesting and everything is a great "new beginning." Wouldn't it be more realistic and constructive to convey the sense of skepticism, or perhaps doubt, that's so often present in science? And couldn't it be done in a positive way so that the public would accept it and see that there's debate and that not everything is "the cure for cancer"?

KOTULAK: That's a good point. I think we in this country did start off being "gee whiz" about scientific discoveries. Don't forget they've been coming at an exponential rate, and they are very exciting. Take genetics, for example. I can remember, early on, going to the Jackson Lab in Bar Harbor to cover the genetic conferences. Scientists would put up on the screen an image of a whole chromosome with a little dark band across it, and everybody got all excited about that. It was interesting stuff, and this was the first time we were seeing something going on in the chromosome. And look where we are today, mixing and matching genes and organisms. A fantastic rate of progress.

Michael Gazzaniga, Ron Kotulak, Colin Blakemore.

But then what happens is that journalists are no longer moved by a single gene discovery (unless it's extraordinary in some way). We evolve, or at least we try to, to the point where we say, What does that gene *do*? What practical use can be made of it? And can it answer some other vital question about biology?

So we keep progressing. The bar keeps being lifted as to what now makes news. One of the classic examples is anti-angiogenesis, which I've been covering for more than twenty years. I remember when Judah Folkman first talked about it and presented his first results. I was very excited. But now that the idea is well known, and taken seriously, we're at the point of "show me." Give me some provocative results here. If you look at anti-angiogenesis today, the results are still not very good. It's a case where there was a lot of promise, but nothing so far has really materialized. That doesn't mean it won't, but it hasn't. We have to put things into focus.

The bar keeps being lifted as to what now makes news.

I think we're becoming more sophisticated at handling these tasks. Part of my job at the *Tribune* is to look at wire-service copy related to science. The national-desk editor will bring over copy that comes in from the AP or Reuters or wherever, and very often it's full of excitement and promise. I try to analyze it and will often call expert sources to help me with their own assessment of what it is. And then I'll tell the editor that either "Yes, this is a good story, let's run it" or "No, let's not bother with it." This is all a learning process for us, and we *have* been learning.

STEPHANIE J. BIRD (MIT): Dr. Blakemore, I was thinking that you probably have some good experience to share with us about how to recruit scientists to participate in public discussions. How do you get people engaged in it? And what kinds of rewards do you point to?

BLAKEMORE: Actually, I think this is a real problem. And until we have mechanisms in place whereby the scientific profession recognizes the importance of this contribution—in promotions and salaries and things like the research-assessment exercise that we have in Britain (which includes this involvement explicitly when rating the quality of research departments)—we'll continue to have a problem.

In the meantime, how do we encourage people? Various mechanisms have been used that are more or less successful. One that's been very successful is "media fellowships," offered by a number of organizations in Britain to give young scientists the chance to work for a few months in a newspaper office or a radio or television company, doing research for them or helping to make a program. Many of those people have gone on to become regular contributors to the media as a result of this experience.

All the research councils in Britain offer training schemes of various sorts to familiarize researchers with the media, usually free of charge.

And there is an organization called the Media Resource Service, which encourages scientists to register their willingness to talk to the media and to list their areas of specialization so that journalists can easily find people to talk to on a particular subject. It has been quite effective.

Overall, the experience is that although there's initial reluctance—for entirely understandable reasons—almost uniformly when people do get involved in communication efforts they enjoy them, and they see the benefits directly to themselves in their work and in the appreciation of their work. And so the rewards are there to be had just from the experience itself.

ILLES: I just have a comment in that regard. One of the things we have to think about is actually introducing into our curricula the means of communicating with the media and conveying science directly to the public, and to include this much earlier in the educational and mentoring process.

BLAKEMORE: I should mention a policy of the Wellcome Trust—the biggest independent medical-funding organization in the world now—when supporting research students. It now requires them to take training in media presentation—a two- or three-day course in media presentation—as part of their scientific training.

Scientists are always reluctant to talk about the things we don't know, maybe because it tends to undermine confidence.

RICHARD BROWN (The Exploratorium): I really appreciate what Colin was saying about imparting the scientific method's critical thinking to people—that this is in many ways more important than transferring a body of knowledge. I wonder if you could take that a step further. Scientists are always reluctant to talk about the things we don't know, maybe because it tends to undermine confidence. But in terms of getting people interested in science—especially young people in becoming scientists—talking about the things we don't know, and how we're trying to find out about them, could be very useful.

This applies especially to neuroscience, where we've heard these enormously important questions being talked about but they are not yet settled—most of the discoveries and breakthroughs lie before us.

BLAKEMORE: My own view is that there are problems in scientists' articulating or even admitting that there are areas of ignorance. We always work, of course, on the edge of ignorance—and it's true that this is a very stimulating way of approaching the presentation of science, to show that it can't immediately answer all questions. But still, we don't quite want to acknowledge the magnitude of the lack of knowledge.

Ronald Duncan, the poet, edited a book some twenty years ago called *The Encyclopedia of Ignorance*, and he managed to persuade a large number of eminent scientists in Britain to write specifically on areas that they didn't know about. But he had great difficulty in recruiting people to that task.

MICHAEL WILLIAMS (Johns Hopkins University): Many of the common reactions that scientists and physicians have to the press are that "They'll misquote me" or "I can't trust them" or "I'm going to be a victim." Often, it's their own fault. They haven't been given the communication skills you mentioned, and they don't understand how the public appreciates things. So they'll say something that to a scientist makes perfect sense, but that may be unintelligible or even misleading to the public. Also, they tend to push aside anything that feels like an emotional response, even though it might be the very thing that would interest the public.

So there are a lot of things inherent in the way we've been taught to think that get in our way of being able to communicate to the public, or to each other. And for the particular area of neuroethics I think we really need to look ahead and overcome that.

BLAKEMORE: Some of those problems are very deep and fundamental. Others are easily dealt with. Presentation skills are very easily learned, really. There are simple rules in dealing with the media to build trust.

Everyone has to face the fact that you're going to get your fingers burnt. There will be times when you say things that are then quoted out of context, or misquoted, or distorted from what you meant to say. There are ways of protecting against that, but they're not foolproof, and one has to recognize that this is going to happen from time to time.

One of the problems is that in communicating with each other, scientists have built a set of conventions that work perfectly well but are entirely different from those that apply in ordinary discourse. Detachment is considered essential, and to be too enthusiastic about your views is negatively perceived by science. One must have a certain coolness about presentation that's symbolized by the use of the passive voice in writing. It implies "That's not really me who's doing these things; they're absolute. They are the truth. They're not contaminated by personal views."

That doesn't work for communicating with the public. You've got to have the passion to show that you're really enthusiastically committed to your own research and be willing to speak in the active voice.

You've got to have the passion to show that you're really enthusiastically committed to your own research.

MELANIE LEITNER (AAAS): Dr. Blakemore talked about media fellowships, and I wanted to mention that there's a long-standing AAAS media fellowship program that enables scientists to learn more about the media by working in that field. But I also had a question for the panel. There's a public perception, specifically in the cloning debate, that the primary motivating force for scientists is curiosity and not concern for the public good—and that this curiosity will even trump concern for the public good. How do we deal with this perception?

GAZZANIGA: That's news to me. Maybe this perception refers to the fact that the public understands so little, for example, about the cloning debate. If you ask the average person to articulate the difference between reproductive and biomedical cloning, or the equivalence between biomedical and therapeutic, you come up with a very low rate of literacy. But the idea of mad scientists just doing things out of curiosity, and the public thinking this way, is a new one on me. To slide off into that sort of thing would of course be terrible, and it's just one more reason for scientists to speak up about their work.

BLAKEMORE: Can I add something to that? I do recognize what you're saying. We get beaten by two sticks. One is that the only thing that drives our work is curiosity—as if curiosity is a bad thing—and the other is that we're only in it to make money. On the animal issue, I've been attacked on both of those fronts: "You're slaughtering animals simply in order to fill your pocket!" and "You're killing animals just to drive your own curious interests, and how can that be morally defensible?" It's very difficult to steer a course between those two.

DAVID SPIEGEL (Stanford University): I have a comment and a question about the social psychology of science-media relations from sort of the other side of the coin.

Dr. Colin Blakemore: To be too enthusiastic about your views is negatively perceived by science.

Dr. Gazzaniga referred to Leon Festinger's brilliant study of what people do when their beliefs are disconfirmed. But there was another wrinkle to his study that affects our relations with the media. When the cult saw that the Earth had not been destroyed and their prophecy was disconfirmed, they changed course. The way they dealt with the cognitive dissonance was to go around trying to convince everybody else that they were right—or at least would eventually be right—after all.

That is a problem because many scientists are reluctant to talk to the media. As we've discussed, they want to be dispassionate. However, there are interest groups that are all too happy to talk to the media and present their so-called scientific viewpoints. And frankly, I can think of some bad examples where the media has fallen hook, line, and sinker for ideas from people who come on as if they're very certain and scientific when they're not. That, to me, is the other side of the coin, and I wonder if you could comment on it.

BLAKEMORE: You're absolutely right. The single-issue groups have exploited the concern of the press to be balanced. At the moment, we're at the stage where media presentations often consist of two people being interviewed, one of whom represents a scientific view, and the other some outlandish, out-of-left-field idea. And these two things are presented as if they're equally credible. Moreover, there's no shortage of articulate individuals, very skilled in public presentation, who can present views that are entirely unrepresentative of conventional science.

There's no shortage of articulate individuals... who can present views that are entirely unrepresentative of conventional science.

KOTULAK: This is a very good point because the media can be misused in that respect, and obviously has been. My view is that part of the job of a science writer is to assess information. It's not as if you want to give "both" sides, but you do want to make sure that the sides you present are legitimate sides. If somebody is way out, our writing has to reflect this, if it's worth mentioning at all.

I don't think that we can just be innocent bystanders and say, "Well, I can now wash my hands of this because I said this side and I said that side." That doesn't do the reader any good, because the reader hasn't got the time or ability that we have to look into all these issues. We have to find out as much as possible, and try to identify the most factual information that we can.

Let me give you an example. One Richard Seed, who's from Chicago, was interviewed by National Public Radio. He said he was going to clone human beings, and they ran that interview.

Reuters then picked it up; they ran a story saying that this guy who claims he's a scientist is going to clone humans. Editors at the Tribune brought the story over to me, and I said it's baloney because right now there's no possible way you can do that. I mean, they had a tough time with Dolly. So there's no way you're going to do this with a human right off the bat. And so we killed the story.

Well, the next day AP picked it up and everybody else did too, so now we had to come back as a second-day story. One of our reporters interviewed Seed; at the same time, we had other reporters going out and doing an investigation of his background. We wrote a story debunking his claims and showing that the guy was not only a lightweight—extremely marginal and on the kooky side—but that he lost his house because he couldn't pay his mortgage and had all kinds of other problems.

Would you believe that the reporter from NPR who did the initial interview called me after all this ran and said, "Ron, I'm glad that somebody finally wrote the truth"? This bothered me considerably—that he would actually run the story knowing that it was a fraud. I hope he learned something from that episode.

This was an exceptional case, though. If you look at the established newspapers and media outlets, they tend not to do that. They may exaggerate a bit sometimes, but they do try to present material in the most honest way they can.

HENRY GREELY (Stanford University): This is a comment that might have a question hidden in it someplace. It was inspired by Colin Blakemore's comment that as people learned more about science, they actually became more concerned. I think that's a real phenomenon and a very deep issue here, much deeper than the problems of media relations.

When members of the public talk to scientists—particularly in neuroscience or biological sciences—they learn things that are unsettling.

Three things are going on in this regard, I believe. First, when members of the public talk to scientists—particularly in neuroscience or biological sciences—they learn things that are unsettling. The discussion yesterday about free will, for example, would be profoundly unsettling to many people, despite the fact that a number of different perspectives were represented. The very idea that scholars are examining, carefully and scientifically, the question of whether or not there is free will—and might actually be able to (gasp!) answer it—is unsettling to people's world-views.

They may feel unsettled because they fear the consequences. Consider, for example, the truth detector. I can imagine people thinking about that and then worrying, "Gee, will my boss use that? Will my spouse use that? The consequences of that could be something that I'm not very happy with."

And then sometimes it's unsettling because of the "yuck" factor. Somebody showed a slide yesterday of a human ear grown on a mouse's back. Well, that's yucky. The human-neuron mouse, if it ever happens, will be yucky to some people. So some science will just be unsettling to people, period.

Second point: As the lay public talks to scientists more, they begin to realize that scientists don't have all the answers. Scientists don't know where the science will go, and they certainly don't know what society will do as a result of the science. And in fact, I think scientists don't really have any more of a clue than anyone else about society's use of the information they provide and how that information will change society.

Well, that's unsettling too, and honest scientists will tell you, "Yes, there are risks. We can't say that there are no risks." So you've got uncertainty and risk, which prompts some in the public to say, "Well, let's not take any risks. Let's slow down. Let's stop. These things could have bad effects. Let's not go there."

Which leads to the third point: The biological

sciences are particularly vulnerable to being stopped or slowed because of public perceptions of risk.

Nobody's thinking very much about any serious regulation of personal computers or video games, because things in those fields are done, for the most part, by private companies—privately and secretly—and so they are essentially immune, or at least protected from, public oversight. But most of biological science is done publicly, in three respects. First, it is publicly funded, which gives the public a hook, as seen by President Bush's use of public-funding limitations to slow embryonic stem cell research. And it's done in the open. Scientists publish—that's what they do—and the press picks it up and people know what's happening.

The second thing is that even where the biological sciences *are* private and *can* be secret—in the pharmaceutical and biotech industries—there's still a public hook in the U.S. with the FDA and in other countries that have something like the FDA. There's an accepted regulatory body that has some power over what even private industry does.

It's understandable that the more people actually hear scientists, the greater their worries may be.

And finally, I think biology is more prone to this because it affects us. It affects humans—human bodies and human minds—things we tend to care about more.

So it's understandable that the more people actually hear scientists, the greater their worries may be. And the greater their power (because of the nature of funding for biological research), the greater their power will be to stop things or slow things down.

BLAKEMORE: I agree with what you say, but I don't think we should be disconcerted by this negative correlation, which is pretty well established. Knowledge of scientific facts in Scandinavian countries and northern Europe in general—Britain and Denmark particularly—is considerably higher than in southern Europe—Italy and Greece, for example. And yet the trust in science is much greater in the south.

But you know, blind faith is not always a good thing. I think the public is much better prepared to accept the reversals of opinion, and difficulties and disappointments, if it doesn't have that blind confidence in the scientific process.

KOTULAK: I'd just like to reemphasize that we have to get across the idea that science is a process. Even at our paper, I've seen some editors who say, "Well, by golly, they said *this*. It was a fact. And now they're saying it's just the opposite." They just get discombobulated about it all. But we have to keep emphasizing—and I try to do that in writing about it—that it is a process. And as you folks deal with media, it's important to emphasize that science is a process of discovery that will never end.

We tend to believe in facts, but they are not *really* facts. A fact today is a different fact tomorrow, and it will be a different fact in the future. They're changeable things. We've seen that throughout the history of science.

GAZZANIGA: I'd like to turn the question around on Hank. If I understood correctly, you're fearful of secrecy in the private sector. But do you see that it has grown to the extent that it now negatively impacts the university, with all these people having joint appointments at, say, Stanford and those companies? I mean, they must have an ethical dilemma every day, as they walk from one of their jobs to the other, as to who says what to whom about which.

So let me make a bold and ridiculous proposal: Do you think it's time for molecular genetics and molecular biology and neurogenetics and neurobiology to get out of the university and go to the research park and let the university grapple with the great unknown questions openly?

GREELY: My honest answer is, How should I know? But I'll go ahead and try to give you a different one. First, though, I'm not afraid of secrecy. I just think that secrecy leads to different sets of issues—different abilities to do things without public oversight. Sometimes that's good; sometimes it's bad.

But to get to your question about the conflicts of interest, they certainly exist. I think they are a significant but not overwhelming problem, though I'm glad I'm not trying to juggle the number of different hats that many of my friends at the medical school wear. I like to see a problem approached by as many different people and places as possible, coming from as many different perspectives as possible. And so I think it's not a bad thing, but a *good* thing, to have people and issues both from a private, for-profit perspective and from a public perspective.

CARL FEINSTEIN (Stanford University): I'm someone who has lived through the MMR/autism tornado that has been going on for the past few years in the United States, and my impression is that it's been even more intense in England. In this situation, there was an initial scientific report that the MMR [measles-mumps-rubella] vaccine might be connected to autism. This news then spread mostly through the Internet, although it's been widely reported in the journals, and it led to a huge and furious outburst from the public, especially from the parents of autistic children.

This reaction has included segments of the public that are generally well informed, and it stepped up into what seemed to be a very antiscientific and generally antivaccination stance as well. So I wonder what people who are experts on the public/scientific interface would have to say about that.

BLAKEMORE: You're certainly right. It's a very, very hot subject in Britain at the moment, even though the clinical and scientific establishment has lined up against the view that there's any risk, as opposed to one maverick clinician (a guy named Wakefield) who claims there's epidemiological evidence for an association.

The worry here, as with other such issues in Britain, is the experience of BSE [mad cow disease], which always lurks in the background. All of us have an image of the agriculture minister, John Gummer, force-feeding a hamburger to his own daughter in front of the television cameras to reassure the British public about the safety of beef—a gesture that subsequently produced considerable embarrassment.

So, for instance, the media in Britain were baying for Tony Blair to say whether or not his own baby, Leo, had received the triple vaccine or not. He tried to hold out against that, I think, with Gummer in mind. Finally, he let slip a few hints that it might have been done, and the media came to the conclusion that certainly it *had* been done, so the issue was closed.

This is a very good example of the need to present a balanced view in responding to a deep and immediate concern of the public, and of what happens if we don't. In some parts of Britain now, the take-up of free vaccination has fallen to about 70 percent, and there's a significant rise in measles outbreaks around the country as a result of the problem. There's a real retreat from vaccination because of this one issue.

RALPH BRAVE (*The Nation* magazine): I'm glad that Ron's an optimist about what's going on with the media in science reporting, but my experience is that it's quite dismal—that the Richard Seed episode you related, in which this fraudulent story was put out there even by some of the top science reporters in the country, is actually typical.

And it's a real problem. I think Ron's up on this panel because he's an exception. But what Art Caplan said yesterday about how people are exposed to these things—for example, learning about DNA through paternity testing on the *Jerry Springer Show*—is more typical of the media trend, which is to try to translate everything into entertainment. This is a real dilemma, and I see it just getting worse and worse.

On the other side, listening to some of the comments from Colin and others about scientists communicating to the public, what I really hear is not so much learning how to communicate but learning public relations. And in a way that's almost noncommunication—learning how not to communicate, how not to enter into a dialogue. As one of the indications of this, the largest growth sector in the National Association of Science Writers' membership is public-affairs science writers—science writers who are now working for corporations, institutions, or universities to help their scientists communicate selectively.

Given the state of media reporting in this area, I don't blame the universities and the scientists wanting to do that, but I think it also stems from a problem that Dr. Gazzaniga pointed out: with the Bayh-Dole Act of 1980 allowing research universities to apply for patents, they now have a very different kind of interest in what gets communicated to the public. So I'm concerned that a whole institutional shift has occurred, though I don't have a solution to it.

When I talk to editors and publishers, it's clear that they don't understand the science, they don't know what's going on, and they simply want whatever's hot, whatever's entertaining. And TV is tremendously problematic in this area. I don't have a solution to this, either.

That's my comment. But I also had a quick question for Dr. Gazzaniga. So far, the President's Council on Bioethics has mainly focused on the cloning issue and stem cells. How do you see it grappling with neuroscience issues? What do you think the first issue in that area will be? And what does your experience so far tell you about how the council will deal with it?

GAZZANIGA: Those are good questions. We had a meeting a month or two ago on what neuroscience is going to look like in 2025 and what some of the issues are likely to be. One that keeps popping up is culpability—reduced responsibility—feeding off this free will issue. Another is cognitive privacy. How is brain imaging going to unlock who we are, and do we really want people to know that? Another issue is one that came up yesterday—therapy/enhancement and the neuroscience domain.

It's a long list, actually. But which issues are going to come up first—which ones are the most pregnant for discussion—are not completely clear to me. I invite everybody to comment on that.

ROBERT LEE HOTZ: I'm a science writer for the *Los Angeles Times* and I'm also on the Board of Directors of the National Association of Science Writers. It's been very interesting listening to this conversation today, but I'm not going to preface my question with a long comment. I will simply ask, In these discussions about the desirability of the public's understanding of science, how do you distinguish between science education and the *marketing* of science? Are we talking about public education here or public relations?

Dr. Michael Gazzaniga: Science is a conservative thing that moves along very carefully.

GAZZANIGA: I think that's a great question. What I was trying to get at earlier was the distinction between scientists and science. Among scientists, there's any number of people who love the limelight and continually promote themselves in one way or another, and we all know who they are. But while scientists come and go, and their ideas come and go, science itself is a conservative thing that moves along very carefully.

Now, as it turns out, on this panel I'm on it's science that's talked about, not scientists. No one there has a particular ax to grind in promoting some new discovery; we're all just trying to evaluate the problem. It's not even a science committee but a group of people with all kinds of different beliefs. Our goal is to understand the data and accurately communicate our sense of it.

KOTULAK: In my optimism, I think we're living in perhaps one of the greatest eras ever known. While the turn of the twentieth century was the start of the golden age of physics, now we've entered the golden age of biology, which is transforming everything we've ever thought of. To me, it's so exciting, it's hard to contain. But at the same time we have to look at the problems that arise, and we have to make sure we're not exaggerating or making false promises. We've already had experience with that sort of thing, and with how it comes back to haunt us.

While the turn of the twentieth century was the start of the golden age of physics, now we've entered the golden age of biology, which is transforming everything we've ever thought of.

Still, I must say that in talking with Colin about what's going on in England and parts of Europe, I think that the media in this country may do a little bit better job of informing the public about science. We tend to be more on the side of science than in an adversarial role, and this promotes better education.

BLAKEMORE: I think we're bound to see ways in which this enterprise is exploited. It's exploited by the media which make science sensational in order to sell newspapers or TV programs. It's exploited increasingly by universities to promote their own image; on EurekAlert and AlphaGalileo—the two Web sites most heavily used by science journalists—you see the hyped copy that universities now deliver being snapped up by the press. It's exploited by some individual scientists to promote their ideas and, in some cases, to earn a lot of money.

But you know, the genie is out of the bottle. We can't stop the process. We live in an information age. The public is hungry for information, and the mechanisms are there for them to get it. It's in our interests that the information they get, on average, is well balanced and accurate. If we don't jump on the bandwagon, the bandwagon will be run by other people and we'll suffer the consequences of public misunderstanding.

The public is hungry for information, and the mechanisms are there for them to get it.

ILLES: All right. I'm going to conclude this very interesting session, and I thank you all very much for participating.

Session

V

Mapping the Future of Neuroethics

Albert R. Jonsen

Emeritus Professor of Ethics and Medicine, University of Washington in Seattle, and Olum Professor of Medical Ethics, University of California, San Francisco

William Mobley

Professor and Chair, Department of Neurology and Neurological Sciences Stanford University

Zach W. Hall

Conference Chair Emeritus Professor and Executive Vice Chancellor, University of California, San Francisco, and former Director of the National Institute of Neurological Disorders and Stroke

JONSEN SUMMARY: Referring to Plato's Republic, Dr. Jonsen began by describing the downside of enhancement—along with the acquisition of wisdom among those of "gold" nature comes greater insight into the world's imperfections and thus diminished personal happiness for them. He then discussed the three types of "mapping" available to bioethicists. "Tectonic" mapping involves the fundamental questions of determinism and reductionism that underlie ethical discourse. "Geographical" mapping largely addresses epistemological questions—that is, how to get around in the "region." Finally, "locale" mapping deals with particular issues, such as treatment/enhancement. Mappers of each type must ultimately interact, often through Socratic dialogue, in order to produce results that are "rich and influential."

ALBERT R. JONSEN: I believe it was Alfred North Whitehead who said that all philosophy is but a footnote to Plato. I've just had the pleasure and daunting task of actually teaching Plato's *Republic* to a group of college freshmen, so I reread the book quite recently. Yesterday, as I was listening to the discussion about enhancement, it struck me that *The Republic* is perhaps the most eloquent and evocative picture of enhancement ever written.

Plato tells us that the "founding myth" of *The Republic* is that all human beings are born with either a bronze nature, a silver nature, or a gold nature—an ancient prevision of modern genetics, at least as some conceive that science. Gold people are selected to become the guardians; they go through arduous training that will eventually lead them to the vision of the good. Then they come back to live in and lead the republic, ruling not by power but by wisdom. Those with silver and bronze natures are destined for lesser tasks, even servitude, in the republic.

We tend to think of such a system today, as complete totalitarianism in accordance with Karl Popper's interpretation of *The Republic* in his book *The Open Society and Its Enemies: The Spell of Plato.* But there's a feature of the republic that we often forget. Plato says that when the guardians return to the world after they've seen this vision of the good, they will inevitably be unhappy because they'll have to rule in a world that is not up to their standards. They will continue to rule out of duty, but they will always yearn to return to contemplation.

So I thought, what a wonderful message for those who favor enhancement. Arthur Caplan has departed, but if he were still here I'd say to him, "Art, beware of enhancement, because the most enhanced people may be the most unhappy. They will have things to do as a result of their 'wisdom,' or even just because of their enhanced skills and knowledge, that could not only be very difficult but also deprive them of joyful lives."

Dr. Albert Jonsen, University of Washington.

I would like to comment on the metaphor of mapping that we've used in this conference. In the last day and a half, it seems that we've drawn maps of three quite different sorts and at three different levels of our topic. The first sort of map, drawn during our first session, was the "tectonic" level of ideas—appropriate enough, since we had an earthquake here in San Francisco last night. That session brought up the questions of determinism and reductionism that return again and again in philosophical reflection because they are insoluble questions. Yet they seem to have great salience for our ways of understanding each other.

One clear instance of that salience was actually mentioned yesterday—the way in which we think about free will. Obviously, this has implications for the way in which we deal with the criminal justice system, for example. But underlying such applications are these fundamental tectonic questions that must continue to be asked, rethought, and refined in terms of the kinds of new challenges that arise.

There's a second level of mapping, which we did not actually discuss much, though there were occasional references to it. I'd call it the geographical level—the hills, mountains, valleys, and waters whose locations we have reason to map in order to learn how to get around in that region. At this level, I think the questions are largely epistemological. That is, how do we think about these issues? What do we do to assure ourselves that certain assertions are reliable when they come from quite different epistemic sources.

When philosophers talk, they make assertions. When scientists talk, they too make assertions. But their respective assertions come from significantly different epistemic sources, or ways of thinking about what one is doing. We need to devote a lot of effort to making linkages between them. Much of Pat Churchland's work has been specifically dedicated to that kind of analysis, and it needs to continue. We have to keep trying, when an assertion coming from science meets an assertion from ethics, to amplify those ideas and reliably put them together in order to move ahead.

The final level of mapping is that of the locale, particularly the "populated regions"—what the English call the built-up areas. These are essentially plat maps and street maps. The mapping of those regions began when we started to talk about the particular issues—the cases, as it were—that have certain boundaries around them: discussions of problems in research with human subjects; discussions about legal accountability in the criminal justice system; discussions about enhancement and treatment; the case we discussed, for example, about deaf children.

In these locales we want to see what particular experiences are, and we add to experience by experiment, data analysis, debate, communication, and the formulation of policy. These issues of mapping the locale are probably what will move ahead most rapidly and precipitously because they are the things that call upon our interests and concerns. Here is where we probably have to start thinking in terms of what the relationships are between the various maps. That is, when do locale questions or geographic questions also have to be thought of in terms of the tectonics, of the deeper questions?

When do locale questions or geographic questions also have to be thought of in terms of the tectonics, of the deeper questions?

To conclude, let me return briefly to ancient Greece, though not to Plato but to Socrates. The Socratic dialogue—basically, people talking and arguing with each other—seems to me to be a genuine representation of the nature of ethical discourse. Ethics really begins with conversation, and it moves on from conversation as people see that disputes are involved. They begin to find out why the disputes are of the sort that they are—whether it's because of commitments to a deep, tectonic kind of question or whether it's a matter of specific points that need to be clarified.

This conference has been a perfect example of that phenomenon; the long discussion periods, which allowed many people to enter in, were in fact a kind of Socratic dialogue and were extremely valuable. So we have begun a dialogue that I hope will turn into something rich and influential.

Summary of the Conference

SUMMARY OF THE SUMMARY: Going session by session and speaker by speaker, Dr. Mobley succinctly reviewed each individual presentation. He also specified the major conclusions derived from each session.

WILLIAM MOBLEY: Zach Hall began by noting a bit of the history that led to the meeting. His charge to participants was to use the meeting to raise questions, to consider the interface between neuroscience, ethics, and philosophy, and to ponder together how advances in neuroscience were likely to affect our conceptions of ourselves and our responsibilities.

Bill Safire reminded us of the Promethean legend and its presumption that God-like qualities might be given to man. He argued that modern breakthroughs in science, and especially neuroscience, have the potential to create similar concerns. This is because neuroscience, in dealing with the brain, touches on personalities—it touches on *us*. It's who we are. And because it is perceived that neuroscience research has the power both to understand the brain and to change the brain, it's not surprising

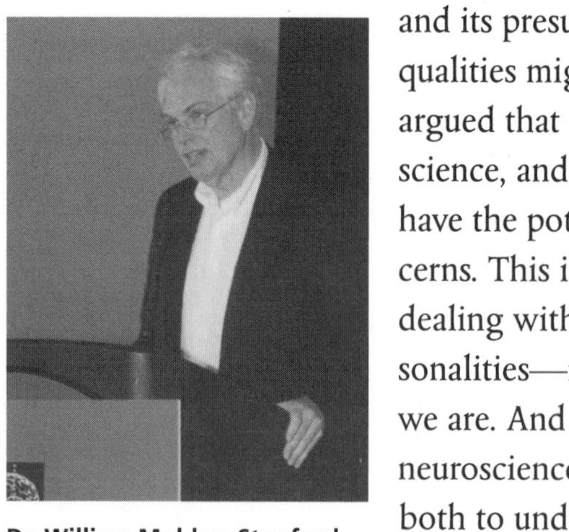

Dr. William Mobley, Stanford University.

that the practice of neuroscience raises concerns.

This new discipline of neuroethics could be the arena in which neuroscience discoveries, and their potential for influencing the well-being of humans, are discussed in terms of what is good and bad, fair and unfair, equitable and inequitable. Potentially, neuroethics could establish rules for participating in brain research, evaluate research claims, determine the relevance of such claims to normal and abnormal brain function, and regulate the use of neuroscience discoveries for diagnosing and treating brain disorders as well as for enhancing the function of normal subjects. The new discipline's domain is heady stuff, literally.

The new discipline's domain is heady stuff, literally.

Session I was devoted to Brain Science and the Self. Anthony Damasio began by talking about ethical behavior. He reminded us that, like all behaviors, it results from the interaction of several neural systems and allows us to optimize our survival and well-being. Because emotion is linked to ethical behavior, failed emotional behavior is the cause of failed ethical decisions and of potentially disastrous social consequences.

Patricia Churchland spoke to us of the self and its capacities. She provided us with a neural model for the self—indeed, a model for self-control—in which various parts have distinct functions. Attention was focused on the "emulator." Informed by perceived emotions and regulated by a conscience, the Churchland emulator judges the potential utility, efficacy, and safety of possible actions before sending appropriate outputs for processing into commands. She argued that there are systematic neurobiological differences between being in control and out of control, and that in time it will be possible to define these two states on the basis of well-characterized neurobiological properties.

Ken Schaffner spoke of the sweeping versus creeping versions of reductionism and determinism, and of the use of these models to conceptualize neuroscience research and assess its implications. The default position here, societally speaking, is one of self-determination, with certain excluding conditions that may exonerate one from responsibility. He noted too that emotions had to be taken more seriously here—that they have to be integrated in some way with our current model. Jonathan Moreno also spoke of free will, defining it as the intelligent release of desire. He noted that modern American medicine assumes that the practice of free will must guide the physician-patient relationship, and he concluded that self-determination is thought to be a realistic moral standard.

The conclusions from the first session, then, were these:

- Ethical and emotional behavior are linked.
- Indeed, emotion and reason may be part of a neural continuum.
- Whatever one calls free will, the brain—as informed by a myriad of internal and external influences—operates to choose among the various desires or choices with which it is presented.
- In this sense, there is self-determination. Or, at the very least, the brain is consciously willing to take credit for the work that it does.

• In order to really understand the brain we need new paradigms. New data will oblige us to throw out old models. In the end, our increased understanding will create, not inhibit, freedom. Less magic is not less interesting.

• Neuroscience discovery will in time cover not just the spectrum that extends from behavior to circuits to cells to molecules; it will eventually extend into the domain of physics.

In Session II we considered Brain Science and Social Policy. Studying the brain offers the seductive promise that by understanding brain function we will gain the ability to make assessments about people, their motivations, their desires, their characteristics. The implications of this promise were discussed and debated.

Memory is not perfect. Dan Schacter reviewed the various causes of memory failure, providing examples of how these shortcomings are creating the desire for new forms of therapy as well as for tests to determine when a memory is false. Regarding the latter, he presented recent neuroimaging findings that showed, in a small group of individuals, the possibility of finding a marker that correlates with false recognition. A question was then raised: Could this be used to reliably distinguish true from false memories? The answer is that it's far too early to tell; the work is in its infancy and the findings are not yet robust. But it's possible that such tests could in time prove reliable.

Bill Winslade spoke of his work with brain-injured people and of the data showing that brain trauma is extremely common among death row inmates. There is currently very little appreciation, however, for the role that clinical neuroscience ought to play in the evaluation of such individuals within the legal system, and he argued that such a role be created and strengthened. Even now, brain science could provide a clearer view of the extent of an individual's brain injury. But with this power comes concerns. At present, they center on the misinterpretation of clinical data—a very reasonable concern. But in the future it may be possible to define the extent to which a person was, or is, out of control or, indeed, to predict a predisposition to criminal behavior. Would such a diagnosis result in incarceration or forced treatment?

Even now, brain science could provide a clearer view of the extent of an individual's brain injury.

Hank Greely spoke about the future of neuroethics, comparing it in several ways with ELSI [the study of the ethical, legal, and social issues created by the Human Genome Project]. The lessons learned from ELSI might well benefit the fledgling effort in neuroethics—apropos of this point, they could illuminate the risks of regulating neuroscience. He also pointed to differences between genetics and neuroscience, noting that while we are not our genes, it was more difficult to say we are not our brains. He concluded by noting the importance of building interdisciplinary teams to define and populate neuroethics.

So conclusions from this session were:

• Before using the new technologies of neuroscience, it will be important to define *normal* and *abnormal.*

• Neuroscience is still a very young discipline. A great deal of additional research will be needed to bring it to the point where it can accurately predict behaviors or define characteristics relevant to most of the issues discussed. We have plenty of reasons to be humble about what we know.

• It is not too early to begin thinking about an effort that would define how to evaluate the applicability of research claims, set guidelines for how they would be used, and translate them into practice.

• The standards we set for defining or predicting characteristics of brain function will evolve with our increasing knowledge of the nervous system.

• Neuroethics must engage all the communities represented at this meeting. As a first objective, I would recommend that we try, if possible, to develop a lexicon that everyone can understand. Commonly defined terms would be especially helpful.

• It will be important to secure both public and private funding to build the discipline of neuroethics.

• The fruits of neuroscience investigation must be the principal drivers of the neuroethics agenda. Let the specifics of what neuroscientists have learned form the questions that solve the real problems, of which there will be many, and not the imaginary ones, of which the number is infinite.

In an engaging lunchtime session, Art Caplan advanced the hypothesis that we should try to improve our brains. He questioned whether there was a practical difference between repairing a deficit and enhancing a normal state.

We're clearly doing the latter already. For example, we modify the environment in which our children are raised and educated in hopes that they'll qualify for the best schools and generally be on a track likely to produce great success. Enhancing children's skills and abilities is a huge emphasis within our families. We are unlikely to find a parent in any culture, in fact, who would not endorse this behavior. The concerns with enhancement are that they could be applied in ways that are unfair, unequal, unforgiving, or unnatural.

Art advanced the idea that these concerns could be dealt with in ways consistent with current standards. In the discussion on the latter point, it was noted that not only are we the products both of cultural and biological evolution, we have been technologically modifying our biological heritage for quite some time. The issue for Art was not whether we should engage in enhancements—he thought that it's highly appropriate—but how best to evaluate the possibilities for doing so and how such enhancements would be regulated. All of this is presumably the domain of neuroethics.

The treatment versus enhancement theme was echoed in Session III: Ethics and the Practice of Brain Science. Steve Hyman spoke of neuropharmacology and the evidence that drugs can evoke long-term changes in the structure and function of neurons. The concerns raised by these data are to some extent balanced by concerns for what happens to neuronal function as a result of *not* treating brain disorders: for the right patients, drugs work, they work well, and they ought to be used. Illustrating his points, he discussed the use of Ritalin for ADHD and pointed to the need for better diagnostic measures, better treatments, more treatments, and better physician training in managing these treatments.

Marilyn Albert spoke of the issues facing

Alzheimer's disease patients and their families. There are no truly effective treatments for Alzheimer's right now. Current ethical issues involve the utility (or lack thereof) of genetic testing, the inability of patients to give informed consent, and the need for more powerful methods for predicting the disorder. This focus will change when effective therapies come on-line; it will then be on presymptomatic diagnosis so that treatment can begin before irreversible changes in function occur.

Erik Parens returned to the issue of enhancement, arguing that the treatment/enhancement distinction might be useful for articulating a basic package of care, critiquing social practice, and affirming natural variation.

A great deal of discussion ensued on this topic. Some viewed it as desirable that enhancements not be used—that is, that there was value in working through a disability. Others came down strongly on the side that enhancement should routinely be used, if available. Steve Hyman reminded us that psychiatry has already dealt with this issue regarding antidepressant medications, and that the benefit of medical psychotherapeutics was not just the ability to treat these conditions more effectively but to actually change the biological model evoked by the word depression. Enhancement changed our view of the biology.

Paul Wolpe sketched a very aggressive view of the future role that technology would play in neurobiology. If this vision comes to pass, we will clearly require a new way of thinking about the interaction of technology and biology, and about how new interventions should be evaluated and regulated.

Conclusions from this session were:

• It's vital that we develop new and effective treatments for neurological and psychiatric disorders. Understanding and treating these disorders should continue to be prominent goals of neuroscience research.

• Moving treatments to the clinic will create new and difficult issues for neuroethics: Who can be treated? What protections for privacy should be in place? Who can give consent? And should the individual have the right to refuse such treatments?

• Neuroethics could help determine how best to move new treatments to the clinic and how to regulate their use.

• As our ability to intervene in brain function increases, it will become increasingly difficult to distinguish treatment from enhancement. We already pursue enhancements and will continue to do so. But neuroethics can play a role in defining under what conditions this practice is harmful to the individual and/or society.

• The fruits of neuroscience investigation should be the principal driver of the agenda for neuroethics.

Don Kennedy spoke last evening of his long-term love affair with science and science policy. He reminded us of animals' extensive behavioral repertoire and noted that many of the characteristics we like to think of as uniquely human are evident in animals. On the issue of free will, he was less concerned with defining it than in carefully documenting the specific neurological phenotypes that could be used to excuse aberrant behaviors. He welcomed the continuing engagement that surrounds the therapeutic use of genetic manipulation, and finally—and in my view, appropriately—placed primary responsibility for ethical decision making in the hands of neuroscience investigators.

In the Session IV we discussed Brain Science and Public Discourse. Colin Blakemore pointed out efforts in the U.K. to educate the public in science, but public confidence has lately been shaken by a series of problems, AIDS and mad cow disease among them. The media linked these problems to scientists and, quite unfairly, placed the blame on them. This crisis has motivated a renewed attempt to engage the public, though in a more general way and marked by genuine discourse: it's not the scientists talking to the public, it's scientists and the public talking together. A number of new initiatives are pursuing this goal, and so far it sounds as though they're going well.

Ron Kotulak talked of his experience as a science reporter. He noted that an important role for the media is in bringing science to the public. It is possible to do this responsibly, and for scientists and journalists to work together as full partners. Indeed, scientists must learn to interact more effectively with the media in order to better communicate the process of science and its benefits.

Mike Gazzaniga discussed his experiences on the President's Council on Bioethics. His recommendations for successful service on such panels included: accept resistance as the norm; take time to educate your colleagues, even those who may not deserve your time and energy (but don't tell them that); and finally, communicate the joy of science to your colleagues and to the public.

The conclusions of this session were that we have responsibilities to educate members of the public, do it well, and be educated by them. This must be a dialogue and one that has to begin immediately. It should be driven by science—driven by the things we know about science and the joy we derive from it.

For me, the overall conclusion of this day and a half is that neuroethics lives. Now is the time to codify the work that we wish this emerging discipline to undertake, to ensure that it engages effectively the scholarly communities that will drive its agenda, and to build the dialogue with the public that will be needed to sustain it. Onward!

Mapping the Future of Neuroethics

SUMMARY: Dr. Hall began by pointing out that decisions are often made unconsciously, based on judgments so familiar as to be automatic, and that a basic task in a new endeavor like neuroethics is to carefully—consciously—examine the issues as an essential precursor to doing the right thing. For the field of neuroethics per se, he suggested that practitioners ensure it develops as a scholarly discipline, that it involve professionals (such as neurologists) who work at the front lines, and that the field not be "left to the experts" but actively include members of the larger society. He also urged that neuroethics focus specifically on two main goals at present: prevention of harm, and protection of the vulnerable. At several points in his talk, Dr. Hall underscored his wish that many bright and motivated young people will enter the field.

Dr. Zach Hall, University of California, San Francisco.

ZACH W. HALL: I want to begin my talk about what we can do next—where neuroethics should go—by saying that this meeting has actually changed my own sense of what neuroethics is about.

We of course have the ability

to make choices, as Dr. Moreno pointed out. It's what we do as human beings, because we have a sense of right and wrong. But we also realize that our ability to make those choices and the range of choices that we have are both biologically and culturally constrained. We recognize as well that because of these constraints, some of us undoubtedly have stronger sensitivities and capacities for thinking about ethical issues than others.

So I was very taken by the comments of one of the participants, Melanie Leitner, who pointed out that habits are those things that we do by and large without conscious choice. We do them sort of automatically, without thinking. And if we want to change things, or if we want to do something new, we have to consciously try to adjust the unconsciousness, so to speak.

I'm very aware of the role of consciousness in changing the ways our nervous system works, and from a particular vantage point. My wife is a musician who spends long hours practicing passages on her instrument, which is the English horn. And we joke that basically what she's doing is putting those passages into her spinal cord. That is, she goes over them very consciously and very slowly, playing each note, hearing each note, being aware of each note. And then she is able, finally, to do it without thinking about it—that is, to have a sort of automatic response. And if she realizes she's making a mistake, then what she has to do is to bring that passage back into consciousness, think about it some more, go through it again and again, remembering each time to press down this finger and not that finger, until she's gotten it right.

It seems to me that this is what we do in many other situations. And in some ways it's what we do in making ethical choices: we first bring matters into consciousness, where we can examine them and be aware of all the factors, and then later act out our choices pretty much automatically. So I really like Patricia Churchland's idea of the emulator, of playing out the different scenarios. And we certainly do that unconsciously. But perhaps we do it most effectively *consciously,* when we actually do think about the various scenarios that may happen—when our reason lets us mindfully play these out. My own view—in agreement, perhaps, with Tony Damasio—is that in the end, when we do play them out, what we listen to is not our rational minds but how we feel about these matters. And it is in our feelings about the various possibilities that ethical decisions are made.

Now I go through all this because I think it's useful for what we're doing as a group, in terms of neuroethics. I would argue that the most important thing we can do is to try to bring into consciousness, and into the consciousness of our society, what we are doing and what the possible consequences are. And then in the end it is the job of our society to decide how it feels about those various consequences—to have the clash of values and show the strong feelings we've observed on a small scale here—regarding social issues.

So, building on this idea that our job is to essentially bring into consciousness things related to research on the brain, what specifically can we foresee for the field of neuroethics, and what do we need to think about?

First, we must obviously consider the development of neuroethics as a scholarly discipline. It's important that our thinking in this field have a base of scholarship, with rich connections to other fields of scholarly inquiry in philosophy, psychology, and the law. It's important to have people who are professionally committed to thinking about these problems—people who will develop with their colleagues a language and traditions that are understood by those in the field and that will allow major points to be thrashed out through reflection, inquiry, scholarship, and learned discussion. And it's important that those who do this be familiar with the nuances of contemporary neuroscience, and also with the rich traditions of philosophical thinking, some of which we have witnessed at this meeting. It is also important for us to understand that we don't start from zero when we think about these problems, but that in fact the greatest minds of our species have been concerned about them for thousands of years.

It's important that our thinking in this field have a base of scholarship, with rich connections to other fields of scholarly inquiry in philosophy, psychology, and the law.

It's also important, as Al [Jonsen] suggested this morning, that there be an arena in which the best methods of philosophy—that is, rigorous, disciplined thinking—be brought to bear on these questions. I see this meeting in that spirit, and I hope it will be an important stimulus to development of a new field of scholarship, with all the accoutrements that we are familiar with—books, publications, journals, scholarly meetings, conferences—in other scholarly fields.

Let me say that, most of all, I hope this field will attract bright young students and professionals. This is a major goal of a meeting like this, and I was enormously pleased to have met several people here who said they have backgrounds in neuroscience and are interested in philosophy, in ethics, and in these particular questions. One young woman came up to me and said, "I want to be the next Pat Churchland." I answered, "Right on! That's what we need."

A second consideration, I believe, should be the importance of neuroethics from a professional point of view. And here we're concerned with some of the issues that Session III raised most prominently. In some ways it's the physicians and scientists who will be at the front lines of many of these questions, carrying out the procedures, prescribing the drugs, doing the experiments. And as a practical matter, it is in their hands that much of the power we've been discussing will lie.

We've seen over the last twenty-five years or so a growing sensitivity to issues such as informed consent and patients' rights. And I think these issues are even more crucial for the brain than other organs, both because of the potentially irreversible nature of the changes that we make and because essential human characteristics are at stake. So education, and heightening the sensitivity of neurosurgeons, psychiatrists, and neuroscientists to these issues, will be tremendously important. We'll have to look to the professional societies—the Society of Neuroscience, the neurology and neurosurgery societies, the psychiatric associations—for some serious assistance here.

We will also look to these professional societies for

providing expert opinion on many of these issues as they come up. We've had some conversations at this meeting about how this might be done, but we've only begun to address this tremendous issue. My hope is that some of the societies, in alliance with The Dana Foundation, will devise effective ways to deploy their expertise.

Neuroethics is not simply a matter for the ethicists or the neuroprofessionals.

That brings me to the third major point here, which I think in some ways is the most important. Neuroethics is not simply a matter for the ethicists or the neuroprofessionals; it must also involve politicians, religious leaders, public policy experts—even columnists from leading newspapers. The issues are simply too important to be left to the scientists.

So the challenge, then, is to figure out how to involve these groups in thinking about neuroethical problems in a proactive way. What happens now, as we know, is that a matter suddenly arises, emotions run high, and the discussion is then dominated by the contingencies and quick reflexive notions of the moment. What's needed instead is some mechanism for identifying, discussing, and marshaling expertise on specific issues before they arise.

We all possess strongly felt prejudices and preconceptions that widely differ, even within this small group. And in the larger society, these issues are going to be magnified even more.

New technology, for instance, usually gets used because it appeals to people. If there's a pill that makes us feel better or perform better, we will almost certainly want to take it because it stimulates our reward centers; it adds to our self-esteem and happiness. And in a society in which the availability of these things depends on the market, one of course sees the potential exploitation of our deep biological needs and the resulting loss of the very control that we seek, to use again Patricia Churchland's very nice phrase. Meanwhile, governments are reluctant to interfere with markets and with people's freedom to do what they want to do.

Think about some of the issues surrounding food. Here we live in a society in which we suffer from a national epidemic of obesity that is having tremendous effects of our health—Type II diabetes, for example. We know what we should do to eat well, and yet we are flooded with advertising for food that isn't good for us but that we consume anyway, and meanwhile we feed fast food to our children in schools. Government doesn't confront these issues directly because of the power of the food industry, basically, and because people like fast foods.

These problems are similar to the issues that neuroethics will be dealing with, without a doubt. That is, people will want certain things resulting from neuroscience that make them feel good in one way or another but that may have deleterious side effects.

With neuroethics, our own particular responsibility is long-term effects on the brain.

So how can neuroethicists help? I think the first thing is to focus on identifying harm. The point is not to say, "You shouldn't take that pill because it's somehow unnatural." The real issue is whether positive harm comes from that pill. Looking at tobacco, for example, it seems to me that in the 1960s the surgeon general's report showing the ill effects of smoking was the first and key point in the whole matter. It simply said, in effect, "Look, this is bad for you." With neuroethics, our own particular responsibility is long-term effects on the brain. We've actually heard some discussion at this meeting that some agents act on the brain most effectively by making long-term alterations.

After identifying issues in which there's harm, our other responsibility is the protection of the weak and disabled, of children, and of other vulnerable populations prone to being exploited. The discussion of deaf children yesterday was a fascinating example of some of the ways in which differing values swirl around these issues. There's obviously no unanimity here. But when it comes to children being mistreated, people who are disabled with psychiatric illness being abused, addicts or others who (for biological or genetic or whatever developmental reasons) are being rendered more vulnerable to particular agents that the rest of us can partake of freely, we need to direct society's attention there. Even though it's not clear what exactly to do about these things, we at least need to identify them and say, "Here is the situation."

Meanwhile, we have to avoid making premature recommendations. One of the dangers we face is that of half-knowledge—of thinking we know what's best when we really do not. Consider, for example, the history of prefrontal lobotomies and psychosurgeries. We do, of course, have to look at all this and *try* to discern what we think is best, but with a large dose of humility. And we need to engage with the larger society on these issues.

Ed Rover (president of The Dana Foundation) asked me a couple of questions that in closing I would pose to you. If we do engage the larger society on these issues, how much effect will we have on it? Conversely, how much effect will the larger society have on our thinking? I hope that the ultimate answer, in both cases, is "a lot."

So we also hope this discussion will continue in various formats. We hope The Dana Foundation will continue to carry the banner of neuroethics and foster it in various settings. We hope the professional societies will pick it up. We hope the scholars who focus on these issues as a primary professional interest will continue to do so.

And most of all, we hope that young people will come in, really concerned from the beginning about these issues, to help us in our task and in enlightening us and our society.

Open Forum and Discussion

JONATHAN MORENO (University of Virginia): I want to underline one topic that Steve Hyman, Marilyn Albert, and a couple of other people have alluded to but that we didn't really talk about. And I want to put another item on the table.

The first one is the importance of the policy issue regarding researchers' use of people with impaired decision-making capacity. This is a "my eyes glaze over" kind of problem. It's boring and uninteresting to most researchers, and they've neglected it. So the result is that state legislatures, for the most part, are going to make the decisions about who counts as a legally authorized representative. And that will have a direct effect on the kind of research that the translational people will be able to do in their institutions.

The other topic—which we haven't touched on—is one that most of us in this room are probably unfamiliar with directly: the military implications of neuroscience. I can assure you that every important article that's published by neuroscientists is vetted by people in the Pentagon or by defense contractors. The biggest investment in neuroscience in the 1950s and 1960s was by national security agencies. Sure, there are a lot of nuts who imagine they're being controlled by the CIA or something. But the fact is that there are significant efforts, funded by our tax dollars in part and not by the nuts, to explore the potential for the management of human beings for military/political ends. So that is something we need to have on the agenda for the future of neuroethics.

STEVEN HYMAN (Harvard University): I think it was well said that when we talk of informed consent, people's eyes glaze over. The fact is that we still don't do a good enough job. As I've often said, informed consent isn't signing a form. That's just a receipt to cover the doctor's ass. Informed consent is an educational process that really has to go on continuously; it doesn't end at the beginning of the experiment. I think that if we don't, as a community, really get this right, we're going to lose the ability to make very important treatments and preventions

Informed consent isn't signing a form. That's just a receipt to cover the doctor's ass.

available to children and to other vulnerable populations—such as those with cognitive and emotional impairments—and that would be an absolute disaster.

My view is that we as a community want to be self-regulating. That is, we don't want to have the heavy and often clumsy hand of legislation telling us how we can conduct clinical trials or any other kinds of experiments. But in order to be self-regulating, we have to be particularly responsible, especially in scrutinizing our own behavior.

The fact is that research is thought of very differently by policymakers and the general public from the way they regard other activities. So, for example, we've worked very hard to make sure that people with mental illnesses can vote. They can have bank accounts; they can have as much self-determination as their condition permits. And yet, when it comes to their ability to consent in research, all of a sudden there's a much more paternalistic attitude. We don't understand this very well yet, but I think it has to do with some of the issues that Colin [Blakemore] was talking about—with the particular distrust of science. These are all important areas for us to be engaged in.

JODI HALPERN (University of California, Berkeley): I also want to add something, prompted by the journalism discussion, that I think is important for the neuroethics agenda. One of my graduate students is looking at journalism's coverage of public health, and we've noticed that if a story only involves an episode, the society looks at the health factors and health decision making involved as caused by individual choice. But if it's a thematic story—which we have little of in this country because of the pressure that journalists are under to tell everything in sound bites—then readers can see how social factors are essentially the basis of health outcomes. So I don't want us to err in neuroethics by being too episodic and not appropriately thematic, especially when we communicate to the public.

It's important that the scientific community be very proactive in trying to convey the values of research, especially when it involves impaired persons.

JENNIFER KULYNYCH (Association of American Medical Colleges): I'm an attorney and I work in research policy in Washington. I wanted to echo the comments about the importance of scientists participating in the education of policymakers and, I'd add, of their participating in the judicial system as well. We've seen the Maryland high court make a ruling this year that in effect would have constricted research with children in any domain that was construed as nontherapeutic. And it has taken a lot of effort by people involved in research policy to try to get that decision cabined off or rolled back. The impact of that ruling would be less direct than regulation: there'd be a chilling effect on that type of research because liability concerns will constrain institutions from conducting it. So I think it's important that the science community be very proactive in trying to convey the values of research, especially when it involves impaired persons.

ZACH HALL: Before we move on to the next question, let me just take the opportunity to say that we appreciate your being here and your speaking up. In organizing this meeting, one of the things we were less successful at was reaching people who are involved in public policy. Part of the problem is that while there are well-defined professional societies and cultures—of neuroscientists and ethicists alike—it's harder to get access to those in the public policy world who are interested in these issues.

STEVE HEILIG (San Francisco Medical Society): I'm coeditor of the *Cambridge Quarterly of Healthcare Ethics,* and before adding a few points let me say that I've been very proud to publish many of the people in this room over the years.

In Professor Winslade's presentation yesterday of the case study of John, there were many striking things for neuroscience. But what struck me the most was that this fellow, who had clearly demonstrated being out of control, was allowed access to loaded firearms—not once but twice. The outcome of his life, and those of others, would have been a lot different if such access were restricted.

The next point results from a conference we put on here in San Francisco just a couple of months ago, chaired by Dr. Phil Lee, regarding the impact of toxics in the environment and in the bodies of children—particularly the impact on their early development. The scientific presentations there were very disturbing, in that toxics are apparently becoming a big contributor to learning disabilities and many other physical disorders.

Another point is that the end users of all the things that neuroscience may develop—therapies, diagnostic tools, and so forth—haven't been addressed very much. In San Francisco, we've found one way to do that—through people who are patient representatives. In the HIV epidemic, for example, this has proven to be very useful. Similarly, for most of the conditions we'll be dealing with in neuroethics, there are expert patient-advocate groups to call on. They'll bring all of the social and public policy issues to you in such abundance you won't be able to stand it!

And finally, on those issues that I mentioned and others, there are huge lobbying industries very interested in preserving the status quo; meanwhile, scientists and ethicists are generally very leery of being activists. But I think that until mainstream science, medicine, and ethics get more politically involved in these things, we'll all be swimming against the tide of social pathology that takes much of what we do here to be of too much academic interest and of not enough interest to the vast majority of people.

Many of the drugs that derive from neuroscience affect the nervous system by definition, and they could be misused by those with hostile intent.

RUTH L. FISCHBACH (Columbia University): Coming from New York, perhaps I'm overly sensitive, but one topic that hasn't come up is bioterrorism. Many of the drugs that derive from neuroscience affect the nervous system by definition, and they could be misused by those with hostile intent. I don't know how, or even whether, this group could approach that concern, but we at least need to be aware of it.

The other point I wanted to make has to do with

the responsibility of investigators. Our technology is moving at such great speed, and we have an imperative to use these new tools, but it really behooves investigators to be responsible. The old mantra of bioethics, "It's not what you can do, rather it's what you should do," is really something we should keep in mind. I've sat on IRBs where we review projects that should never have come up before us because they violate the rights of people or are just plain harmful. I'm happy that this conference has been keeping these concerns in mind, and we here can heighten responsibility by going back to our institutions and setting the direction through positive example.

ERIK PARENS (The Hastings Center): I wasn't there at the beginning of the ELSI [ethical, legal, and social issues] movement, but based on what I've observed over the past decade or so my guess is that in the beginning nobody tried very hard to specify what they meant by *ethics*. I thank you, Zach [Hall], for trying to do that.

Now let me see if I heard you correctly. It seems to me there are many legitimately different ways to understand what ethics is. Some people would say that it's an attempt to understand what we ought to do. I thought I heard you say, rather clearly, that this is not how you understand the purpose of neuroethics but that, instead, a far more appropriate and productive purpose is to identify what is harmful. And given your tobacco example, I understand you to mean what is harmful to people's bodies.

You also said that the young, the disabled, and other vulnerable populations ought to be protected. You didn't say from what, but I assume you meant from harm. So my question—to the group, I suppose—is this: Is there agreement that by *neuroethics* we mean a conversation aimed at protecting people from bodily harm—bodily being broadly understood—and not about this other thing of what ought we to *do*? Is there agreement about that or not?

HALL: I have no idea if there's agreement, but I certainly gave my own view. It seems to me that our grounds for being able to say what people ought to do in these situations is very limited. But I thought we could surely agree on the two things that I mentioned. So at the very least, we can bring information to bear, try to point out the consequences, play out alternative scenarios, and then say: "Here's what happens if you do this, as far as we can tell. And by the way, understand that this child has no opportunity to say no here."

It's a lot easier to agree about harmful consequences to bodies than it is about harmful consequences to the society.

PARENS: But the point I'm trying to make is that, as I tried to say yesterday, I myself am fundamentally interested in consequences. Some are easier to talk about, and get agreement on, than others. It's a lot easier to agree about harmful consequences to bodies than it is about harmful consequences to the society, for example. We might want to rule out, up front, conversation about what we ought to do with respect to the society. I don't need to tell you what I think about that.

HALL: What we suggest has to be evidence-based, I

think. And that's the big advantage of talking about what's harmful in individuals: based on what we know now, or based on our evidence from this series of experiments, we think people will face this and that consequence. But to say, "This could destroy the moral fabric of our society," for example, is an unprovable statement that's much more dangerous and provocative. At that broad level, there's almost certain to be disagreement no matter what is being referred to.

DAVID McGONIGLE (University of California, San Francisco): My subjective feeling about this conference is that we've made the most advances, and had the liveliest discussions, when there has been application of a great body of ethical knowledge to novel discoveries and novel questions in neuroscience. And I was wondering if you thought that one of the greatest challenges to this nascent field of neuroethics will be when there is direct conflict between novel neuroscientific findings and that body of ethical knowledge.

WILLIAM MOBLEY: I tried to speak to this. Maybe this is my medical background coming out, but what I've found most useful and illuminating at the conference has been the cases. When the John case was discussed, for example, it provided a whole set of thoughts and feelings directly relevant to John and his problems. It's important that we use cases to figure out what the problems of neuroethics really will be, and the more concrete they are—especially when linked to what we can actually do, or what we'll likely be able to do in the future, with neuroscience—the better.

ALBERT R. JONSEN: Let me also address that question. The history of bioethics is a history of crisis. It came into being over several critical issues, the first being experimentation with human subjects. A vast world of research with human subjects had developed over one hundred years, with little recognition of ethical parameters. As some scandalous events took place—such as when the details of the Tuskegee syphilis study came to light—public attention became focused on the ethical issues involved. The reaction, more instinctive than philosophical, was that the rights of certain persons were being violated only to serve others. But the attempt to analyze this fundamental problem—of the relationship between the welfare of individuals in research and the good to society that comes out of research—called for some fairly serious and honest thinking that did draw on the great body of philosophical knowledge.

One paper in particular was extraordinarily influential in formulating the way in which we, and the National Commission for Protection of Human Subjects, began to devise regulations and organize research. That paper was written by Hans Jonas, a very distinguished philosopher at the New School in New York, who had not addressed many practical matters before but was nevertheless asked to consider this question of research with human subjects. Jonas was able to draw on a very rich philosophical tradition in order to analyze this new question of how we deal with the attainment of knowledge at the price of possibly violating the rights of individuals, and the result of his inquiry was a very important step.

Now, I'm not saying that neuroethics is going to need a crisis in order to get started. I hope it doesn't happen, but the fact is that this is frequently part of the history, and it helps. So when we do have a crisis, the question becomes: Do you respond simply at the gut level, with an "Oh my God " and a preoccupation with the yuck factor, or do you respond by bringing to bear an intellectual tradition that can analyze it and put it in focus. That's a crucial question for neuroethics, as it always has been for bioethics.

HILLEL BRAUDE (University of Chicago): My comments are related to the previous few, although formulated a little differently. In Session I, Brain Science and the Self, Al Jonsen said that one of the purposes of this conference was to join the two continents of philosophy and neuroscience. But while I think that we have explored both shores of these continents during the last few days in a very rich and productive fashion, I'm not all that convinced they've been bridged.

Most of the discussion has been about the "neuro" aspect of neuroethics—its promises and perils. Meanwhile, the "ethics" aspect has been limited to reaction to what can be done in the face of these technologies, which will progress whether we would like it or not. Ethics as an "ought" has been somewhat avoided.

In any case, as has come out in the last few minutes here, whose ethics are we talking about? There is no real consensus on how to use these technologies.

So I don't have any simple solutions for how to make neuroethics really ethical with a capital E. But I do think that we need to bring to this discussion as great a diversity of voices as possible.

I don't have any simple solutions for how to make neuroethics really ethical with a capital E.

I was at the ASBH—the American Society for Bioethics and Humanities—for the first time last year, and apparently there's a discussion there about whether the society should endorse particular opinions. Perhaps twenty years down the line, the neuroethics community will be having the same kind of debate on whether or not to present a unified face to the public. And I think our decision, like ASBH's, should be that we represent diverse ethical voices—that there should not be a single type of neuroethics. One of the temptations of neuroethics' neuro aspect is to reduce ethical questions to essentially neurological ones, and I think that should be avoided.

JONSEN: I'm reminded by your reference to the ASBH that in the absence of a crisis, one way to begin these sorts of discussions very fruitfully might be to ensure that somebody puts together and sponsors a workshop on neuroethics at the American Association of Bioethics. And similarly at the Society for Neuroscience. It happened also, in the history of bioethics, that in the early years the Hastings Center was very active in getting workshops sponsored at places like the AAAS [American Association for the Advancement of Science] meetings. And those were very valuable for building and maintaining a wider understanding that there are issues worth exploring here.

HALL: The Society for Neuroscience actually had a social policy session last fall. Several of the speakers from that session are here, and I'm going to recognize one of them now.

HENRY GREELY (Stanford University): I want to start an argument: I don't think *neuroethics* is a very good term for this field. I don't think it's an accurate word for it. And that requires me to tell you what I think the field *is*.

To me, what we should be studying—what we should be worrying about—are the consequences for our societies, our cultures, and our lives of the new information in neuroscience. In other words, how it's going to affect how people live—individually, in families, and in societies. We should be trying to predict what those social consequences are going to be, whether they're going to be—and here are a contentious couple of words—*good or bad,* and how, if at all, we can intervene to try to increase the benefits and decrease the costs.

So, for example, I think that deciding what rules the judiciary should apply in determining whether or not to admit brain images in criminal cases, how primary education should be changed, and what privacy laws should apply to neuroimages are all issues in this field, whatever the field is. They are not issues that would normally be viewed as ethical issues, except in the very broadest sense of the "e" word ethics. I'd prefer we used a term something like "social consequences of neuroscience."

I have another reason for not liking the "e" word. And this may be a little more contentious and may even sound somewhat parochial. I think that the people who explore this domain will have to come from lots of different disciplines and fields, and that no one field can dominate. The neuroscientists can't dominate it, and the philosophers can't dominate it. But philosophers, I've found, have a tendency to view the word *ethics* as their own private property, particularly in a field like neuroscience where issues like free will draw philosophers the way a light draws moths. Having the very name of the field reinforce the possible idea of a primacy of philosophy is a mistake. I'm afraid this is a doomed argument because I don't have a better word. Neuroethics *sounds* great.

HALL: How about neurolaw?

GREELY: But it shouldn't just be neurolaw. I don't think lawyers should dominate this any more than anyone else.

We have to apply a very, very broad definition of ethics in order to make the term neuroethics sensible.

HALL: I was being facetious.

GREELY: Of course, we *will* dominate it. But I don't think we should. This really has to be a broadly interdisciplinary activity, and it troubles me that I don't have a better word than *neuroethics.* It's catchy, it's memorable, and it was coined by the chairman of The Dana Foundation—all of which argue for its continuing existence. But I think that in using it we need to keep in mind that we have to apply a very, very broad definition of *ethics* in order to make the term *neuroethics* sensible— a definition that takes into account all the social consequences of the really exciting and somewhat scary new area of science that it complements.

JONSEN: Hank, bioethics has been broadly interdisciplinary.

STEPHANIE J. BIRD (MIT): Hank Greely and Art Caplan started to talk about the topic of ELSI yesterday, and that was kind of short-circuited because we didn't have the time. But from my experience and perspective I would say that the existence of ELSI really was valuable because it did identify the ethical, legal, and social issues—the societal context of the science—as being so fundamental. And that's why I believe it would be essential to follow that precedent in other kinds of neuroscience funding.

I think that designating a component, or set-aside, in particular for each grant has the downside of focusing on the issue that the particular grant raises. So having a broader source of funding that allows for looking at the larger issues of neuroscience/neuroethics would be critically important to actually making this effort go forward. Foundations as sources is a really good idea, but it would say something more for our society as a whole if the national funding agencies agreed to do it. Either way, we'd need a direct mechanism for getting the results of the work translated into practical social policy.

PATRICIA CHURCHLAND (University of California, San Diego): I just got a little note from Hank Greely saying, "But you're a philosopher I like."

The thing that's right about what Hank said is that ordinary folk—down-to-earth common sense people like fishermen, carpenters, and nurses—often have much more to say on ethical issues, and more reasonable views, than people who are supposed to do ethics professionally. So I think it's important for us to have some humility.

But I think that the term *neuroethics* isn't so bad, because part of what I'd like to know in the long run is how the brain does ethics. I mean, what is the brain doing when it does moral reasoning? It's not as though morality, or values, have a kind of transcendent existence somewhere. We don't really believe anymore that they're handed down from a divinity or anything like that. So it has to come out of brains and from the ways that brains interact with other brains. I'd hope this is something we will ultimately address, and the social implications of our understanding it may turn out to be kind of interesting as well.

When you do neuroethics, or when you do neuroscience that has ethical implications, you need to have collaborative consultation—you need to work *together*.

MARY ELLEN MICHEL (NIH): I really enjoyed a slide yesterday noting that one of the obligations of the field is to make these difficult issues available to the meanest intelligence. I work for the government, so that really spoke to me. [Laughter] And what it really tells me is that we shouldn't just build another ivory tower where there are a lot of really smart people in philosophy and ethics and neuroscience or whatever, all studying neuroethics. I think that there's this kind of instinct to do that—to try to surround oneself with people who are intellectually stimulating. But we really need to make some of these arguments available to the "meanest intelligence"—to reach out and get a public debate going—so that the debate is really informed rather than just the mental exercises of a new, cool field that has meetings in beautiful places and insulates itself.

HALL: Let me just say that I think we need both. It's important to have people involved who are knowledgeable about what other people have thought on these issues and who bring careful reasoning to bear—who are able to proceed in a logical and disciplined way. But that's not enough, and the other conversation needs to take place too.

The expertise part is a sort of reservoir, it seems to me, from which the larger conversation can draw, in much the same way that the neuroscientists' knowledge of how the brain works, or their lack of knowledge about it—and knowing where that boundary is—is also a reservoir of information. Still, the important conversation is the larger one. In my view, everything depends on that conversation not only taking place but being as informed and as civilized and as proactive—that is, occurring not just in the heat of the moment. Admittedly, as Al [Jonsen] says, it may take the heat of the moment to get people to sit down and talk. But even so, we need to have as much preparation for those moments as possible.

WILLIAM J. WINSLADE (University of Texas Medical Branch): I'd like to agree with Hank about the importance of collaborative, interdisciplinary work. But I disagree about changing the name; I *like* the term *neuroethics.* What's needed is not just a name but a clarification of the concept, a statement that elaborates on what neuroethics is that would include the kinds of things people here have been saying. In other words, it's open to everybody who wants to address the ethical, legal, and social issues, and it's not an exclusive preserve for philosophers and would-be philosophers.

Along with the idea of collaboration, I'd like to say a word in favor of funding *not* being like the ELSI model, which has a lot of inherent problems. I've served as a reviewer of numerous ELSI projects, and it seemed to me that especially in the early years the ELSI projects were "over here" and the genetics projects were "over there." There was very little collaboration, or communication, or serious discussion about the issues.

I think that when somebody does a neuroscience research project there should be an ethics aspect that is not just reserved for neuroscientists to hold forth about or for philosophers to carp about. It seems to me that when you do neuroethics, or when you do neuroscience that has ethical implications, you need to have collaborative consultation—you need to work *together.* And that's the way funders ought to address it. Now, this doesn't mean you can't have separate projects as well, but it does seem to me that it would be much better than what I see as the flaws of the ELSI model.

JONSEN: We've been using the word *field* of neuroethics. We've not used the word *discipline.* A similar question arose in bioethics. It was thought of as a field, and only later as a discipline. The difference in my mind between the two is that a discipline has a structure to it that a field doesn't have. I mean, a field is a big place where you can run around and jump and play, and so you come out and jump and play with your ideas. A discipline has, well, some discipline.

The structure began to appear in bioethics when teaching began—when people started to teach courses in bioethics and had to create textbooks, those execrable volumes with certain structured ways of going at problems. Of course, other things—commissions, study groups, and the like—were happening too. But it all has to start somewhere, and the teaching part was it. For neuroethics, though, we may not be ready to start that process now.

A point about the ELSI project—a lot of really interesting work has been done over the years. But my problem with the project is that the most important players haven't played. That is, the leading geneticists and microbiologists have not been part of that program, even though it was practically designed to be theirs. And some people of prominence who have nominally been part of it haven't really played either.

I can give you an example. At a conference in Europe a few years ago one of the leading American molecular biologists, who always said that this ethics stuff is wonderful, came as a speaker. He gave the opening talk, and then he and his family spent three days touring, not attending a single meeting. He came back only at the end to give the summary. It was marvelous creativity, but a . . .

HALL: . . . *lack* of ethics, I would say. [Laughter]

JONSEN: I think that many of the scientists, the people who really are the main players, have either been bored by ELSI or haven't known what they were to do in it. So while they've talked the talk, they didn't walk the walk.

The final question that I raise, though, is: What should a scientist *do* with all the talk? Paul Berg did something very important some years ago when he recognized that molecular biology had reached a point at which there had to be a serious examination of the recombinant DNA research that was going on. So the scientists had a meeting, and they created what turned out to be a kind of self-imposed moratorium. That was a very important act, though it's rare that scientists have made any contribution like that at all. And that's a big problem. How do you get people who are interested in doing good science—that's where their lives are, that's where their objectives are—turn around and say to their colleagues, "Oh, there's something new for us to talk about—namely, the ethical implications of what we're doing."

I would not welcome the field of neuroethics becoming a hermetic NIH-funded industry.

MICHAEL STRYKER (University of California, San Francisco): I don't know much about neuroethics, I suppose, but I know what I like! And I would not welcome the field of neuroethics becoming a hermetic NIH-funded industry.

What's interesting and important for discussion—much of which has been alluded to here—are topics in which our increasing knowledge of neuroscience changes our view of the nature of man. And in that discussion, the voices we should probably listen to most are completely unrepresented at a meeting like this—the voices of the artistic community. These are the people who write imaginative literature, the people who write of ideas about the mind and ideas about the body and how they have influenced our conceptions of who we are. This is a very hard perspective to incorporate, but it really does have tremendous influence on the public, on the law, and probably on most social thought about neuroscience.

So we should deal with the social issues that Hank mentioned, which I think are very, very important ones. And scientists and philosophers and groups like this one gathered here should have a voice and contribute to these kinds of discussions. But on the larger issues—"What are we?" "What is the nature of man?"—that intrigue us so much and probably got most of us working in neuroscience to begin with, the dialogue really needs contributions from the artistic community at least as much as it needs them from the people who define the field of ethics.

MARY MAHOWALD (University of Chicago): I agree with you. Some of the concepts we've been keying in on during these discussions are actually broader than moral philosophy. To talk about the meaning of the self and problems of identity and problems of autonomy gets us into the meta-ethical—they're beyond the area that bioethicists generally work in. So we do need a broader term. How about something like *neurohumanities*? That at least extends beyond the narrowness of both philosophy and ethics.

WILLIAM HURLBUT (Stanford University): Adding to what you said, I think we should be very careful to understand that this is an interface of the most significant kind, that here science is not just talking about physiological-function stuff. We're talking about the mystery of the psychophysical unity of being that is the human mind, or the human existence. In that sense, I think your comment about bringing in artistic perspectives is very apropos.

But I also think we should extend it. I mean, where do we get our concept of what our significance is? Pat Churchland said a very interesting thing—that she'd like to study how the brain does ethics. In the original Greek, *ethics* means "habit," "custom," or "character," which obviously flow forward from the most foundational things that give us a sense of significance.

Where do we find our significance? Traditionally, we've sought it in some notion of what is called in religion "spiritual anthropology." So the mystery is: What is this concept of ethics that we have in our being from the foundation? Is it in the service of life? If so, what kind of service? Does it just serve survival, in which case maybe it's a very self-serving thing? Does it differ among individuals? If it does, do we see differences among racial groups or ethnic groups or geographic groups? If we do, would we then turn eventually toward a technology of moral alteration or moral enhancement?

The point is that there are very, very deep issues here. They go far deeper than a kind of functional service to the society to keep it from using things in a wrong way that's damaging to the physical health. The question underneath all of this is: What is life for? What gives life its deep significance? And could we, in the process of discovering these things, change our image or change even our capacity to feel these issues of significance?

When the contraceptive pill was introduced… this was probably the first time that tens of millions of people were using a medicine to cure something that was not a disease.

Just to give one very brief example: When the contraceptive pill was introduced—I was a medical student at the time—it struck me that this was probably the first time that tens of millions of people were using a medicine to cure something that was not a disease. It was a major paradigm shift, at least to my mind. And how did it affect us? Well, we've gotten used to it, but in the meantime a lot of people's lives have been changed in ways they didn't expect. There was a sexual revolution. There was a lot of deferred pregnancy, and a lot of unhappy barrenness subsequent to it, and so forth.

That's a trivial example compared with what we're entering into now. We're entering into something of great—of ultimate—significance. And in that sense we need to turn to our traditions that have told us what makes life significant in the first place. In the end, neuroethics is the big question. It's the final ethical question. And we need to ask about what kind of a mind we have, what makes it meaningful.

We have to be very careful not to denigrate the degree to which we get our sense of meaning out of our natural existence. That's because a lot of our existence is unconscious. It's under the surface. It bubbles up without our understanding it. We may want sex, but we get children, in the natural environment. And if you take those two and dissociate them, you have to be very careful that you don't walk yourself right off the stage of the drama of your deepest significance.

Einstein said that the most incomprehensible thing about the universe is that it's comprehensible. I'm sure he meant the mathematically comprehensible. It would be interesting to know if the same thing applies to the moral meaning of the universe. And if so, does it depend on a particular construction of biology?

Call it what you will, the neuroethics train has left the station.

FRANCIS HARPER (The Dana Foundation): In the six years that I've worked with The Dana Foundation, not being a scientist I've had to learn patience with the progress of science. And now I discover I'm going to have to learn patience with the progress of ethics. And with the progress of neuroethics. Still, I applaud the effort to establish a field, to map terrain, to build a mentorship for students, to create scholarly publications.

Call it what you will, the neuroethics train

has left the station. The debate is on, and the question is: While you create a field to carry forward the debate, who is dealing with it in the public arena and who is making the decisions? If you take Dr. De Sousa's view that morals are about sex, and ethics is about money, I suppose it's all right for this debate to take place in the U.S. Congress. However, I really think—that was my only attempt at a joke—it's imperative for, and even incumbent on, responsible neuroscientists and ethicists to formulate the short-term plan, to put the finger in the dike, if you will.

What are you going to do when the questions arise now? How are you going to address the issues that are before us now? Thanks to neuroscience, the questions about who we are as humans are going to continue to be answered. For many, many years we're going to have a lot of time to debate this in a structured, organized, well-rewarded, and well-recognized field. But meanwhile I think the responsible thing to do is for all of us to figure out, in the short term, how to collaboratively help society make some of these important decisions.

HALL: That's a challenge, and an appropriate note on which to end. Thank you all for coming.

Appendix I

Participant Biographical Information

Marilyn S. Albert, Ph.D. is Professor of Psychiatry and Neurology at the Harvard Medical School. Since 1981, she has been the Director of the Gerontology Research Unit at Massachusetts General Hospital, and in 1999 was appointed Director of the Harvard-Mahoney Neuroscience Institute at the Harvard Medical School. She has authored over 200 academic publications and with husband Dr. Guy McKhann, recently saw the publication of their book for the general public on the aging brain entitled *Keep Your Brain Young*.

Colin Blakemore, Ph.D., Sc.D., FMedSci, FRS is a Waynflete Professor of Physiology at the University of Oxford and is also Director of the Oxford Centre for Cognitive Neuroscience. He has been President and is now Chairman of the British Association for the Advancement of Science; he was President of the British Neuroscience Association from 1997 to 2000, and is now President of the Physiological Society. He is also Chief Executive of the European Dana Alliance for the Brain. Dr. Blakemore is a Fellow of the Royal Society, the Academy of Medical Sciences and the Institute of Biology.

ARTHUR CAPLAN, PH.D. is currently the Emanuel and Robert Hart Chair for Bioethics and the Director of the Center for Bioethics at the University of Pennsylvania in Philadelphia. At Penn, he is also a Professor in the Department of Philosophy and in the Department of Psychiatry in the School of Medicine. Dr. Caplan is the author or editor of more than 20 books and over 500 papers in refereed journals of medicine, science, philosophy, bioethics and health policy. He is the recipient of many awards and honors including the McGovern Medal of the American Medical Writers Association, Person of the Year 2001 from USA Today, and six honorary degrees from colleges and medical schools.

PATRICIA SMITH CHURCHLAND, PH.D. is UC President's Professor of Philosophy and Chair of the Philosophy Department at the University of California, San Diego. She is also an Adjunct Professor at the Salk Institute in La Jolla. Her research targets neurophilosophy—the interface between traditional philosophical questions and developments in neuroscience. Her most recent book is *Brain-Wise: Studies in Neurophilosophy*.

ANTONIO DAMASIO, M.D., PH.D., the Van Allen Professor and Head of Neurology at the University of Iowa and Adjunct Professor at The Salk Institute, has had a major influence on the understanding of the neural basis of emotion, language and memory, and consciousness. The laboratories that he and wife Hanna Damasio created at the University of Iowa are a leading center for the investigation of brain and mind. His books *Descartes' Error: Emotion, Reason and the Human Brain* and *The Feeling of What Happens* are taught in universities worldwide. His new book, *Looking For Spinoza,* will be published by Harcourt in 2003.

MICHAEL S. GAZZANIGA, PH.D. currently serves as Editor-in-Chief of the *Journal of Cognitive Neurosciences,* and heads the McDonnell Summer Institute in Cognitive Neuroscience. He is also the founder of the Cognitive Neuroscience Society, and started the Center for Cognitive Neuroscience at Dartmouth College. In 1997, Dr. Gazzaniga was elected to the American Academy of Arts and Sciences, and in 2002 was appointed to the President's Council on Bioethics.

HENRY T. GREELY, J.D. is the C. Wendell and Edith M. Carlsmith Professor of Law and a professor, by courtesy, of genetics at Stanford University. His specialties are health law and policy and legal and social issues arising from advances in the biosciences. He chairs the steering committee of the Stanford University Center for Biomedical Ethics; directs the Center for Law and the Biosciences; and co-directs the Stanford Program on Genetics, Ethics, and Society. He serves on the California Advisory Committee on Human Cloning and on the North American Committee of the Human Genome Diversity Project, whose ethics subcommittee he chairs.

ZACH W. HALL, PH.D. is the President and CEO of EnVivo Pharmaceuticals, a biotechnology company whose purpose is to discover and develop pharmaceuticals for central nervous system diseases. Prior to that he was the Executive Vice Chancellor and Professor of Physiology at the University of California, San Francisco, where his major administrative responsibility was the development of the 43-acre basic science campus at Mission Bay. From 1994 until 1997, Dr. Hall served as Director of the National Institute of Neurological Disorders and Stroke (NINDS) and the National Institutes of Health, and prior to that, as the Chair of the Department of Physiology at UCSF.

STEVEN HYMAN, M.D. is Provost of Harvard University. From 1996 to 2001, he served as Director of the National Institute of Mental Health (NIMH) at the National Institutes of Health. Among his honors, Dr. Hyman is a member of the Institute of Medicine of the National Academy of Sciences. He serves on a number of scientific advisory boards, including the Howard Hughes Medical Institute, the Riken Brain Sciences Institute in Japan, and the Max Planck Institute for Psychiatry in Germany.

JUDY ILLES, PH.D. is a Senior Research Scholar at the Stanford Center for Biomedical Ethics, with a joint appointment between the Departments of Medicine and Radiology. She co-founded the Stanford Brain Research Center (SBRC) in 1998, and served as the SBRC's first Executive Director between 1998 and 2001. Dr. Illes is the author of *The Strategic Grant-Seeker: Conceptualizing Fundable Research in the Brain and Behavioral Sciences,* and special guest editor of a forthcoming issue of *Brain and Cognition: "Ethical Challenges in Advanced Neuroimaging."*

ALBERT R. JONSEN, PH.D. earned his Doctorate of Religious Studies from Yale University. He initiated the bioethics program at the School of Medicine, University of California, San Francisco in 1971 and became Chairman of the Department of Medical History and Ethics, University of Washington in 1987, becoming Emeritus in 1998. He is a member of the Institute of Medicine, National Academy of Science. Among his books are *Clinical Ethics, The New Ethics, and the Old Medicine, The Birth of Bioethics* and *A Short History of Medical Ethics.* He is currently Visiting Professor Emeritus, Stanford Medical School.

DONALD KENNEDY, PH.D., a biologist by training, is currently Editor-in-Chief of *Science,* the journal of the American Association for the Advancement of Science. During his career, Dr. Kennedy has served as Provost, then for twelve years as President of Stanford University. He is the author of *Academic Duty,* a book discussing some of the challenges facing American institutions of higher education. Dr. Kennedy was elected to the National Academy of Sciences in 1972 and also serves on the Board of Trustees of the Carnegie Endowment for International Peace.

BARBARA A. KOENIG, PH.D., an anthropologist who studies contemporary biomedicine, is Executive Director of the Center for Biomedical Ethics, and Associate Professor, Department of Medicine at Stanford University. Her latest project examines the ethical and policy implications of evolving knowledge in the genetics and neurobiology of addiction. In 1999 Koenig was named a Faculty Scholar of the Open Society Institute's "Project on Death in America." In 2002-03 she will be a Fellow at the Stanford Humanities Center.

RON KOTULAK has been a *Chicago Tribune* Science writer since 1963. He received the 1994 Pulitzer Prize for explanatory journalism for two related series on brain research: "Unraveling the Mysteries of the Brain" and "Roots of Violence." Kotulak is past president of the National Association of Science Writers and received the American Diabetes Association's C. Everett Koop Medal for Health Promotion and Awareness.

BERNARD LO, M.D. is Professor of Medicine and Director of the Program in Medical Ethics at the University of California, San Francisco. He is a member of the Institute of Medicine (IOM) and chairs the IOM Board on Health Sciences Policy. Dr. Lo was a member of the National Bioethics Advisory Commission that issued reports on stem cell research and research on mental disorders that may affect decision-making capacity.

WILLIAM MOBLEY, M.D., PH.D. is Professor and Chair of the Department of Neurology and Neurological Sciences at Stanford University. He also serves as co-Director of the Stanford Brain Research Institute. He is the recipient of both the Zenith Award and the Temple Award from the Alzheimer's Association and is a Fellow of the Royal College of Physicians. Dr. Mobley serves as Editor of *Neurobiology of Disease* and as President of the Association of University Professors of Neurology.

JONATHAN D. MORENO, PH.D. is the director of one of the nation's largest bioethics centers at the University of Virginia. Among his books is *Undue Risk: Secret State Experiments on Humans,* which was nominated for the Los Angeles Times Book Prize. His other books include *Deciding Together: Bioethics and Moral Consensus, Ethics in Clinical Practice,* and *Arguing Euthanasia.*

ERIK PARENS, PH.D. is the Associate for the Philosophical Studies at The Hastings Center. He also teaches in Vassar College's program in Science, Technology, and Society. His current research explores the ethical and social implications of medical technologies aimed at shaping ourselves and our children. He is the editor of *Enhancing Human Traits* and *Prenatal Testing and Disability Rights.*

WILLIAM SAFIRE, winner of the 1978 Pulitzer Prize for distinguished commentary, joined *The New York Times* in April 1973 as a political columnist. He also writes a Sunday column, "On Language," which has appeared in *The New York Times Magazine* since 1979. This column on grammar, usage, and etymology has led to the publication of 12 books and made him the most widely read writer on the English language. In 1993, he became a member of the Board of Directors of The Dana Foundation. In 1998 he was elected Vice Chairman, and on the death of his lifelong friend, David J. Mahoney, was elected Chairman in 2000.

DANIEL L. SCHACTER, PH.D. became Professor of Psychology at Harvard University in 1991, and has been Chair of the department since 1995. Schacter studies the psychological and biological aspects of human memory and amnesia, emphasizing the distinction between conscious and nonconscious forms of memory, and the mechanisms involved in memory distortion and forgetting. His most recent book, *The Seven Sins of Memory: How the Mind Forgets and Remembers*, was named a *New York Times Book Review* Notable Book of the Year for 2001.

KENNETH F. SCHAFFNER, M.D., PH.D. is the University Professor of Medical Humanities and Professor of Philosophy at the George Washington University. His most recent book is *Discovery and Explanation in Biology and Medicine.* He has been a Guggenheim Fellow and has published extensively in philosophical and medical journals on ethical and conceptual issues in science and medicine. He is currently completing his next book, *Behaving: What's Genetic and What's Not, and Why Should We Care?*

WILLIAM J. WINSLADE, PH.D., J.D. is a James Wade Rockwell Professor of Philosophy in Medicine, and is a member of the Institute for the Medical Humanities at the University of Texas Medical Branch, Galveston, Texas. He is also Distinguished Visiting Professor of Law at the University of Houston Health Law and Policy Institute. Winslade is a member of the California Bar Association and is a licensed Research Psychoanalyst in California. He is the author of *Confronting Traumatic Brain Injury: Devastation, Hope and Healing.*

PAUL ROOT WOLPE, PH.D. is a Senior Fellow of the Center of Bioethics at the University of Pennsylvania, where he holds appointments in the Departments of Psychiatry and Sociology. He is Director of the Program in Psychiatry and Ethics at Penn, and is a Senior Fellow of the Leonard Davis Institute for Health Economics. Dr. Wolpe also serves as the first Chief of Bioethics for the National Aeronautics and Space Administration (NASA). He is the author of the textbook *Sexuality and Gender in Society* and the end-of-life guide *In the Winter of Life*

Appendix II

Conference Planning Committee

ZACH W. HALL, PH.D., CONFERENCE CHAIR
Emeritus Professor and Executive Vice Chancellor, University of California, San Francisco, and past director of the National Institute of Neurological Disorders and Stroke

HOWARD FIELDS, M.D., PH.D.
Professor and Vice Chair of Neurology, University of California, San Francisco

FRANCIS HARPER
Executive Vice President, The Dana Foundation

JUDY ILLES, PH.D.
Senior Research Scholar, Stanford University Center for Biomedical Studies

ALBERT R. JONSEN, PH.D.
Professor Emeritus of Ethics in Medicine, University of Washington and Visiting Professor of Ethics in Medicine, Stanford University

BARBARA A. KOENIG, PH.D.
Associate Professor of Medicine and Executive Director, Stanford University Center for Biomedical Ethics

BERNARD LO, M.D.
Professor of Medicine and Director, Program in Medical Ethics, University of California, San Francisco

WILLIAM MOBLEY, M.D., PH.D.
Professor and Chair, Department of Neurology and Neurological Sciences, Stanford University

Conference Organizing Team

ANNE FOOTER, M.S.
Assistant Director, Stanford University Center for Biomedical Ethics

JOYCE PRASAD
Administrative Associate, Stanford University Center for Biomedical Ethics

ROCHELLE LEE
Stanford University Undergraduate and Dana Foundation Intern

Index

A

B

C

E

F

G

H

I

J

K

L

M

N

O

P

Q

R

S

T

Y

Z